Tsunami Science and Engineering II

I0063582

Tsunami Science and Engineering II

Special Issue Editor

Valentin Heller

MDPI • Basel • Beijing • Wuhan • Barcelona • Belgrade

MDPI

Special Issue Editor
Valentin Heller
The University of Nottingham
UK

Editorial Office
MDPI
St. Alban-Anlage 66
4052 Basel, Switzerland

This is a reprint of articles from the Special Issue published online in the open access journal *Journal of Marine Science and Engineering* (ISSN 2077-1312) from 2018 to 2019 (available at: https://www.mdpi.com/journal/jmse/special_issues/bz_tsunami_science_engineering_II).

For citation purposes, cite each article independently as indicated on the article page online and as indicated below:

LastName, A.A.; LastName, B.B.; LastName, C.C. Article Title. *Journal Name* **Year**, *Article Number, Page Range.*

ISBN 978-3-03921-672-7 (Pbk)
ISBN 978-3-03921-673-4 (PDF)

Cover image courtesy of Denis Istrati.

© 2019 by the authors. Articles in this book are Open Access and distributed under the Creative Commons Attribution (CC BY) license, which allows users to download, copy and build upon published articles, as long as the author and publisher are properly credited, which ensures maximum dissemination and a wider impact of our publications.

The book as a whole is distributed by MDPI under the terms and conditions of the Creative Commons license CC BY-NC-ND.

Contents

About the Special Issue Editor

Valentin Heller is currently an Assistant Professor in Hydraulics in the Department of Civil Engineering at the University of Nottingham. His research mainly concerns experimental and computational fluid dynamics into landslide- and iceberg-tsunamis, fluid–structure interactions, and scale effects. Before moving to the University of Nottingham, Dr Heller held one of the prestigious Imperial College London Research Fellowships, a Research Fellowship at the University of Southampton, and a postdoctoral position at ETH Zurich. He obtained his Ph.D. at ETH Zurich for his work on landslide-tsunamis in 2007. He is an editor of Landslides, an Editorial Board Member of the Journal of Marine Science and Engineering, run three Special Issues, and has been a technical reviewer for nearly 40 journals to date. Dr Heller was also honoured with the Harold Jan Schoemaker Award from IAHR in 2013 for his review on scale effects and with the Maggia Price 2004 for his diploma thesis on ski jump hydraulics.

Journal of
Marine Science and Engineering

MDPI

Editorial

Tsunami Science and Engineering II

Valentin Heller

Environmental Fluid Mechanics and Geoprocesses Research Group, Faculty of Engineering,
University of Nottingham, Nottingham NG7 2RD, UK; Valentin.Heller@nottingham.ac.uk;
Tel.: +44(0)-115-748-6049

Received: 11 September 2019; Accepted: 11 September 2019; Published: 13 September 2019

Keywords: earthquake-tsunamis; landslide-generated impulse waves; landslide-tsunamis; long wave run-up; numerical modelling; physical modelling; seismic tsunamis; tsunami hazard assessment and mitigation; tsunami loading on structures

Earthquake-tsunamis, including the 2004 Indian Ocean Tsunami, with approximately 227,898 casualties, and the 2011 Tōhoku Tsunami in Japan, with 18,550 people missing or dead [1], serve as tragic reminders that such waves pose a major natural hazard to human beings. Landslide-tsunamis, including the 1958 Lituya Bay case, may exceed 150 m in height and similar waves generated in lakes and reservoirs may overtop dams and cause significant devastation downstream, such as in the 1963 Vajont case with nearly 2,000 casualties [1].

In January 2018, I was invited by the editorial office of the *Journal of Marine Science and Engineering* to act as guest editor of the special issue *Tsunami Science and Engineering II* to collect articles about tsunamis with the aim to repeat the success of the special issue *Tsunami Science and Engineering* [2]. I was very keen on this project aimed at representing a wide range of high-level contributions to capture the recent increase in research activity in the field of tsunamis due to a series of recent catastrophes such as the 2018 Java and Sumatra volcanic-eruption-triggered landslide-tsunami, the 2018 Sulawesi earthquake-triggered underwater landslide-tsunami, the 2017 Greenland landslide-tsunami, the 2015 Chile earthquake-tsunami and the 2011 Tōhoku earthquake-tsunami, amongst others [1].

This book includes nine excellent contributions [3–11] to this special issue published between 2018 and 2019. The overall aim of the collection is to improve modelling and mitigate the destruction of tsunamis and the negative effects they have on us and our environment. The articles cover a wide range of topics around tsunamis, and reflect scientific efforts and engineering approaches in this challenging and exciting research field.

The order of the nine articles [3–11] in this book follows the three tsunami phases: generation, propagation and their effects. The first article from Perez del Postigo et al. [3] focuses on tsunami generation and propagation. The large magnitude of some tsunamis has been justified with a dual-source mechanism, for example as a combination of an earthquake- with a landslide-tsunami. Perez del Postigo et al. [3] is one of the first study investigating such a dual-source mechanism. They developed a unique experimental set-up in a 20 m flume to reproduce a coupled-source tsunami generation by an underwater fault rupture followed by a submarine landslide. One of their key-findings is that for a coupled-source scenario, the generated wave is crest led, followed by a trough of smaller amplitude decreasing steadily as it propagates along the flume.

The following two articles from Tan et al. [4] and Tessema et al. [5] involve the entire process of subaerial landslide-tsunami generation, propagation and runup. This is challenging, both from a numerical (typically requiring several models) and physical modelling point of view given that multiple physical processes at different length and time scales need to be modelled. Tan et al. [4] present a numerical landslide-tsunami hazard assessment technique, illustrated with hypothetical scenarios at Es Vedrà, offshore Ibiza, involving the site-specific bathymetry and topography. The violent wave generation process is modelled with the meshless Lagrangian method smoothed particle hydrodynamics and the simulations are continued with the less computational expensive non-hydrostatic non-linear shallow water equations code SWASH. The up to 133 m high tsunamis decay relatively fast with distance from Es Vedrà to 14.2 m offshore Cala d'Hort (3 km from the source) with a maximum run-up height of around 21.5 m. Nearly all numerical and physical model studies into landslide-tsunamis released the slide in the longitudinal direction of the flume or basin, which does not well represent slide impacts in narrow reservoirs in lateral direction. This shortage motivated Tessema et al. [5] to investigate landslides impacting a reservoir in lateral direction in a 1:190 laboratory scale model for a typical dam of ≈60 m in height. They derived new empirical equations for the dam overtopping volume in a function of the slide volume, slide release height, slide impact velocity, still-water depth and upstream dam face slope. They further compare the measured overtopping volumes with a two-dimensional (2D) case reported in the technical literature and highlight some discrepancies supporting the need for their new results.

The next two contributions from Kian et al. [6] and Santos et al. [7] focus on wave propagation and inundation. Frequency dispersion can be very important for tsunami propagation. Kian et al. [6] apply the tsunami model NAMI DANCE, based on the computational efficient nonlinear shallow-water equations, to a range of problems from the technical literature. They show that for certain conditions of grid size, time step and water depth, the model is well capable of capturing small physical dispersion. Kian et al. [6] conclude that their model represents an acceptable alternative to the more computational demanding nonlinear Boussinesq-type equations models if dispersion is small. Santos et al. [7] provide a very detailed record and reconstruction of the 1755 Lisbon earthquake, tsunami and fire in downtown Lisbon. To achieve this, they combine detailed analysis of historical data with tsunami modelling based on the TUNAMI-N2 code and a field survey. Santos et al. [7] found that the number of residences in downtown Lisbon decreased from 26,200 to 6000–8800 due to this catastrophe, partially due to 1000 fatalities. Further, the buildings were mainly destroyed by fires rather than the earthquake, as was previously believed, followed by tsunami inundation of up to 200 m into seafront streets and squares.

Evers and Boes [8] focus on impulse wave (tsunami) runup based on an experimental dataset containing 359 runup heights from the technical literature. This extensive dataset is compared to a range of existing empirical and analytical runup equations. Based on this analysis, Evers and Boes [8] propose a superior semi-empirical prediction equation representing the dataset with an overall ±20% scatter for a wide range of slope and wave conditions.

The last three articles from Istrati et al. [9] and Ghodoosipour et al. [10,11] investigate the impact of tsunamis on bridges and pipelines, respectively, motivated by recent tsunamis damaging and/or destroying such critical infrastructure. Istrati et al. [9] investigate tsunami impact on a bridge with open girders to decipher the tsunami overtopping process and the associated demand on the bridge and its structural components. They find that the maximum horizontal and vertical forces do not always occur simultaneously and that the application of these forces at the centre of gravity of the deck does not yield conservative estimates of the uplift demand in individual connections as offshore connections have to withstand the largest uplift among all connections. They propose "tsunami demand diagrams" as visual representations of the complex variation of the tsunami loading and demand a paradigm shift in the assessment of tsunami risk to coastal bridges to include the distribution of this load to individual structural components, rather than only the total tsunami loading.

Both Ghodoosipour et al. [10] and Ghodoosipour et al. [11] rely on the identical experiments with Ghodoosipour et al. [10] carefully introducing into the dam-break flow characteristics and the influence of the presence of pipelines on flow conditions for a wide range of Froude numbers. The experiments involved both dry and wet beds to assess the influence of different impoundment depths and still water levels on the hydrodynamic features. Ghodoosipour et al. [11] investigate the hydrodynamic forces and moments these tsunami-like bores [10] exert on pipes. Based on their results, Ghodoosipour et al. [11] found resistance coefficients in the range of 1.0 to 3.5 and lift coefficients in the range of 0.5 to 3.0, which are very valuable for the design of pipelines located in tsunami-prone areas.

These brief summaries illustrate the wide range of relevant and fascinating topics covered in this book. I would like to thank all Authors for their excellent articles and for contributing to the success of *Tsunami Science and Engineering II*. I hope that these articles will help to mitigate the negative effects of tsunamis and inspire many future research activities in this important research field.

Valentin Heller: Guest Editor "Tsunami Science and Engineering II"

Conflicts of Interest: The author declares no conflict of interest.

References

1. Wikipedia. List of tsunamis. 2019. Available online: https://en.wikipedia.org/wiki/List_of_tsunamis (accessed on 10 September 2019).
2. Heller, V. (Ed.) Tsunami science and engineering. 2016. Available online: https://www.mdpi.com/books/pdfview/book/209 (accessed on 10 September 2019).
3. Perez del Postigo, N.; Raby, A.; Whittaker, C.; Boulton, S.J. Parametric study of tsunamis generated by earthquakes and landslides. *J. Mar. Sci. Eng.* **2019**, *7*, 154. [CrossRef]
4. Tan, H.; Ruffini, G.; Heller, V.; Chen, S. A numerical landslide-tsunami hazard assessment technique applied on hypothetical scenarios at Es Vedrà, offshore Ibiza. *J. Mar. Sci. Eng.* **2018**, *6*, 111. [CrossRef]
5. Tessema, N.N.; Sigtryggsdóttir, F.G.; Lia, L.; Jabir, A.K. Case study of dam overtopping from waves generated by landslides impinging perpendicular to a reservoir's longitudinal axis. *J. Mar. Sci. Eng.* **2019**, *7*, 221. [CrossRef]
6. Kian, R.; Horrillo, J.; Zaytsev, A.; Yalciner, A.C. Capturing physical dispersion using a nonlinear shallow water model. *J. Mar. Sci. Eng.* **2018**, *6*, 84. [CrossRef]
7. Santos, A.; Correia, M.; Loureiro, C.; Fernandes, P.; da Costa, N.M. The historical reconstruction of the 1755 earthquake and tsunami in downtown Lisbon, Portugal. *J. Mar. Sci. Eng.* **2019**, *7*, 208. [CrossRef]
8. Evers, F.M.; Boes, R.M. Impulse wave runup on steep to vertical slopes. *J. Mar. Sci. Eng.* **2019**, *7*, 8. [CrossRef]
9. Istrati, D.; Buckle, I.; Lomonaco, P.; Yim, S. Deciphering the tsunami wave impact and associated connection forces in open-girder coastal bridges. *J. Mar. Sci. Eng.* **2018**, *6*, 148. [CrossRef]
10. Ghodoosipour, B.; Stolle, J.; Nistor, I.; Mohammadian, A.; Goseberg, N. Experimental study on extreme hydrodynamic loading on pipelines. Part 1: Flow hydrodynamics. *J. Mar. Sci. Eng.* **2019**, *7*, 251. [CrossRef]
11. Ghodoosipour, B.; Stolle, J.; Nistor, I.; Mohammadian, A.; Goseberg, N. Experimental study on extreme hydrodynamic loading on pipelines. Part 2: Induced force analysis. *J. Mar. Sci. Eng.* **2019**, *7*, 262. [CrossRef]

© 2019 by the author. Licensee MDPI, Basel, Switzerland. This article is an open access article distributed under the terms and conditions of the Creative Commons Attribution (CC BY) license (http://creativecommons.org/licenses/by/4.0/).

Journal of
Marine Science and Engineering

MDPI

Article

Parametric Study of Tsunamis Generated by Earthquakes and Landslides

Natalia Perez del Postigo Prieto [1,*], Alison Raby [1], Colin Whittaker [2] and Sarah J. Boulton [3]

[1] School of Engineering, University of Plymouth, Plymouth PL4 8AA, UK; alison.raby@plymouth.ac.uk
[2] Department of Civil and Environmental Engineering, The University of Auckland,
  Auckland 1142, New Zealand; c.whittaker@auckland.ac.nz
[3] Centre for Research in Earth Sciences, School of Geography, Earth and Environmental Sciences,
  University of Plymouth, Plymouth PL4 8AA, UK; sarah.boulton@plymouth.ac.uk
* Correspondence: natalia.perezdelpostigoprieto@plymouth.ac.uk

Received: 8 April 2019; Accepted: 10 May 2019; Published: 17 May 2019

Abstract: Tsunami generation and propagation mechanisms need to be clearly understood in order to inform predictive models and improve coastal community preparedness. Physical experiments, supported by mathematical models, can potentially provide valuable input data for standard predictive models of tsunami generation and propagation. A unique experimental set-up has been developed to reproduce a coupled-source tsunami generation mechanism: a two-dimensional underwater fault rupture followed by a submarine landslide. The test rig was located in a 20 m flume in the COAST laboratory at the University of Plymouth. The aim of the experiments is to provide quality data for developing a parametrisation of the initial conditions for tsunami generation processes which are triggered by a dual-source. During the test programme, the water depth and the landslide density were varied. The position of the landslide model was tracked and the free surface elevation of the water body was measured. Hence the generated wave characteristics were determined. For a coupled-source scenario, the generated wave is crest led, followed by a trough of smaller amplitude decreasing steadily as it propagates along the flume. The crest amplitude was shown to be influenced by the fault rupture displacement scale, whereas the trough was influenced by the landslide's relative density.

Keywords: tsunami generation; submarine landslide; fault rupture; physical modelling; coupled-source

---

1. Introduction

  Recently tsunamis have become the focus of renewed international research efforts after high profile events claimed thousands of lives in countries such as Indonesia, Japan, Thailand and Chile. It has been long known that coastal areas that are tectonically active are prone to tsunami inundations, as earthquakes generated by fault rupture are one of the most common tsunami triggering mechanisms. Usually, larger earthquake magnitudes ($M_w > 7.5$), cause more destructive tsunamis, which tend to be associated to thrust faulting [1]. However, strike-slip faults have also been associated with some of the most devastating events in history, e.g., the recent Palu $M_w = 7.5$ earthquake and tsunami in September 2018. Furthermore, lower magnitude earthquakes are also capable of generating significant tsunamis and have been termed 'tsunami earthquakes' [2], e.g., Newfoundland earthquake ($M_w = 7.2$) and tsunami in 1929, known as *'The Grand Banks Tsunami'*, which broke 12 submarine telegraph cables and is considered the Canada's worst natural disaster [3]; and the Papua New Guinea earthquake ($M_w = 7.1$) 1998, which generated a tsunami which struck the Northwest coast of Papua New Guinea with waves up to 10 m high and causing 15 m runup [4]. The aftermath of the Papua New Guinea tsunami was investigated by Tappin et al. [5], who found evidence of an offshore slump near Sissano Lagoon. Synolakis et al. [6] concurred with Tappin et al. [5] that a submarine landslide also contributed to the tsunami generation process, owing to the relatively moderate earthquake, the wave-height and run-up distributions, and wave arrival time [5]. Therefore, it is possible that coupled-source tsunamis, where submarine landslides are triggered by co-seismic shaking, might be more common than is realised. However, it is extremely difficult to get real time evidence/data of a submarine landslide owing to the unpredictably and accessibility of such events, although bathymetric surveys might help to locate areas prone to mass failure. High-resolution observational data are not generally feasible to obtain due to the complex location of occurrence for submarine landslides.

Consequently, the mechanics and kinematics of submarine landslides have yet to be studied in detail. Investigations of tsunamis generated by earthquakes are usually undertaken with numerical modelling of real data such as satellite, buoy and seismic records (e.g., [7–10]). Experimental modelling of underwater earthquakes is complex due to the difficulty of scaling the fault slip and geometry to water depth in a laboratory facility. Physical models of fault ruptures have been represented either by vertical motions of a plate or triangular moveable pistons, tested in two dimensional flume tanks [11–13]. The vast majority of previous investigations studying landslide generated tsunamis focused on subaerial landslides, both for physical [14–17], and numerical modelling [18,19], for which submarine landslide studies are also common [20–22]. There are also some experimental investigations of submarine landslide-induced tsunamis [23–25]. Usually, physical models of landslides consist of solid blocks of various geometries (e.g., triangular, semi-elliptic) [24,26,27] or a granular mass (i.e., [28,29]).

This investigation presents initial results from a novel coupled-source tsunami generation mechanism, which comprises a fault rupture and a fully-submarine landslide. Physical modelling of this dual mechanism will facilitate the investigation of complex interactions between the two tsunami sources and the affected water body.

2. Experimental Investigation

The test rig replicates two geological events: an underwater fault rupture followed by a submarine landslide. The fault rupture consists of a sudden uplift (up-thrust) of a plate, which is controlled by an electric actuator. A broad set of configurations were tested during each experimental campaign. Firstly, the fault rupture was tested in four different configurations: a horizontal plate uplift (HU), inclined slope uplift (IU), an inclined slope uplift with moving landslide (IU-ML) and an inclined slope uplift with fixed landslide (IU-FL). Moreover, different fault rupture uplift distances were performed: 0.06 m; 0.04 m; 0.02 m; 0.01 m and no displacement, in which case the landslide slid from the top of the slope under gravity (ML). Four water depths were used: 0.32 m; 0.27 m; 0.22 m and 0.165 m. Throughout the various experimental campaigns, different landslide models, as described in Section 2.3, were tested on the IU configuration. The measured parameters were: uplift displacement time-history, landslide position, and free surface elevation time-history.

2.1. Experimental Setup

The tsunami generation experiments were conducted in a 20 m long flume with a 0.6 × 0.6 m section (Figure 1). The test rig, which comprises a square plate (fault rupture model), bed slopes and landslide models, partially sits on a grill where water is normally recirculated when the flume is being used for current generation in its normal operation (Figure 2). The grill conveniently permits circulation of water when the uplift of the fault plate is performed preventing unwanted suction effects. The wave generation was restricted to one-way wave propagation by a backwall. A false floor was placed along the propagation area to provide sufficient clearance between the uplifted plate and the flume floor and to maintain a constant depth (Figure 2). The left most region of the test rig was obscured by the flume support (in blue, Figure 1), thus the camera field of view of the slope was restricted to an area of 0.175 m × 0.6 m within the glass walled region of the flume (Figure 2).

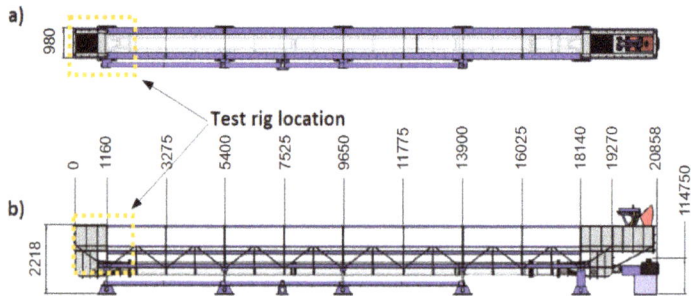

Figure 1. Sketch of the 20 m two dimensional flume of the Coast Laboratory (University of Plymouth). The yellow dashed lines mark the area where the test rig was placed. (**a**) Plan view. (**b**) Cross-section.

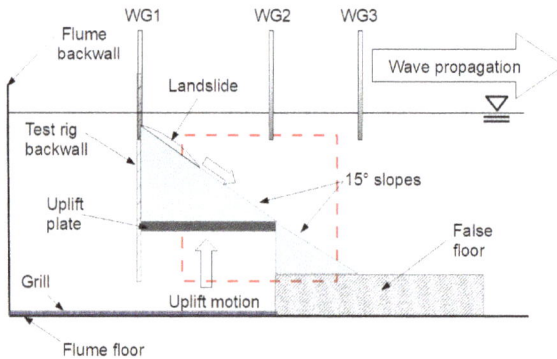

Figure 2. Schematic diagram (not to scale) of the test rig showing the fault rupture (uplift plate) and landslide model. WG indicates resistance wave gauges used to measure the surface elevation changes. The dashed red lines show the camera field of view.

Resistance wave gauges (WG) were used to measure the surface elevation changes along the centreline of the flume. They consisted of a pair of 275 mm long 1.5 mm diameter stainless steel wires spaced 12.5 mm apart. In general, the gauges were approximately 0.3 m apart, but different gauge configurations between runs allowed to extend the surface elevation measurements further along the flume. Table 1 shows the wave gauge locations along the flume, where $x^* = 0$ is the uplift back wall. Wave gauges in the near field were aligned at the changes in bathymetry and the initial location of the landslide (Figure 2). Note that WG1 location in this study corresponds to the $x/h = 0$ location in [30], and WG2 to $x/h = b$.

Table 1. WG locations WG1-WG13 in meters.

	WG1	WG2	WG3	WG4	WG5	WG6	WG7	WG8	WG9	WG10	WG11	WG12	WG13
x^*(m)	0	0.64	0.87	1.25	1.657	2.657	2.907	3.157	3.457	3.757	4.047	4.357	4.657

2.2. Fault Rupture Mechanism

Figure 3 shows the fault rupture model before it was located in the flume. The square plate, which represents the uplifted seafloor, is attached to two vertical side plates that slide up and down the flume walls on guide rails. The plate dimensions are 0.61 m side length (similar to Hammack [30]) by 0.6 m width (corresponding to the flume width). The vertical side plates are joined by a cross-brace where the actuator is attached to perform the uplifting motion. All plates are made of aluminium, the horizontal plate is 5 mm thick and the remaining elements are 10 mm thick to ensure they do not flex during the uplift motion. Stiffening bars were added to the plate to prevent it from bending due to the large uplift force. Two fault ruptures were replicated: a horizontal plate uplift and an inclined plate uplift. The addition of an inclined plane to the test rig enables the landslide to be released, which then slides down the slope, guided on PTFE runners. The benchmark configuration [31] of a 15° slope and a rigid semi-elliptical landslide model were adopted. The uplift motion performed by the actuator in the current investigation aims to reproduce the half-sine displacement time history of Hammack [30] with a maximum displacement of 60 mm, as later explained in Section 3.1.

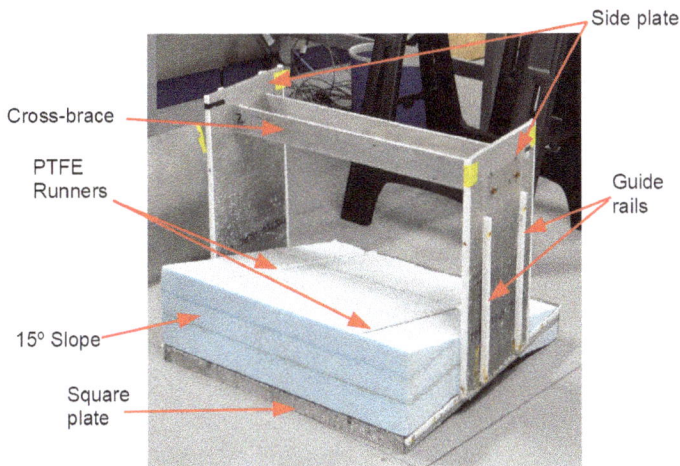

Figure 3. Photograph showing the fault rupture model.

2.3. Landslide Models

2.3.1. Solid Block Landslide

The landslide was represented by a semi-elliptical rigid body. Three landslide models with different thicknesses and densities were tested (Figure 4).

Models ML2 (8.7 kg) and ML2+ (10.6 kg) were hollow with a sealable bottom hatch to enable lead to be inserted to vary the mass of the model whilst keeping the volume constant. A bolt was fixed to the rear of the models as part of the release mechanism. The landslide was automatically released once the plate reached its maximum uplift displacement, at which point the bolt slid through a keyhole on the backwall. The landslide then moved downslope under gravity.

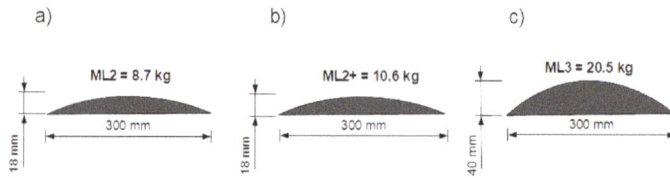

Figure 4. Cross-sections sketch of landslide models: (**a**) ML2, (**b**) ML2+, and (**c**) ML3 and ML3B (same dimensions, divided in slices (Figure 5)).

2.3.2. Modular Landslide

Model ML3 aimed to represent a granular landslide (Figure 5). The main feature of this model was its ability to change shape during its motion. This model consisted of twenty slices that were tested in differently grouped configurations, which were secured together with tape (Figure 5). To achieve the 20.5 kg weight, the landslide was constructed of a combination of stainless steel and aluminium. Two stainless steel rods were inserted through the uplift plate and the slope, and fixed onto the flume floor. When the uplift was performed, both plate and slope moved through the rods, leaving them flush once the required vertical distance was reached, releasing the landslide. This landslide had the advantage of being able to be tested in various configurations. In this paper, two sets of tests are presented for the modular landslide. The tests using the landslide as a whole solid block (all the slices attached together) are referred to as ML3, whilst the tests with independent sliding slices/blocks are referred to as ML3B.

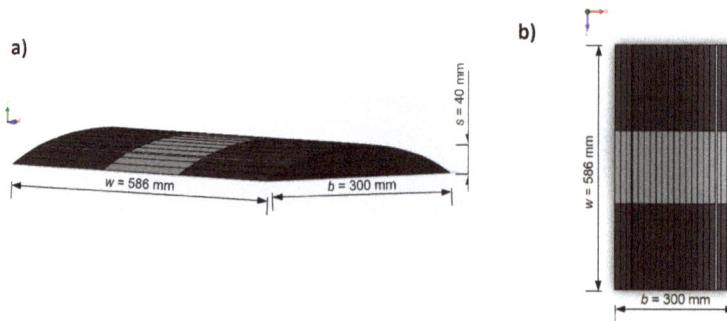

Figure 5. (**a**) ML3 landslide model geometry and dimensions. (**b**) Top view of ML3. Light grey slices are made of solid stainless steel machined bars. Black slices are made of aluminium machined bars.

3. Results

During the experimental programme, in addition to varying the water depth and the landslide size and density, four different fault rupture uplift distances were tested. Four non-dimensional parameters are used to report results: relative density $\gamma = \rho_l/\rho$, where ρ_l is the landslide density, ρ is the density of the water; landslide Froude number $Fr = v/\sqrt{gh}$, where v is the landslide maximum velocity, g is the gravitational acceleration and h is the water depth; normalised wave height η/h, where η is the water surface elevation; and the displacement size scale ζ/h, where ζ is the vertical uplift displacement distance. The results were used to investigate the effects of both sources on the generated wave. Tests were first performed with each source separately before investigating the influence on the generated wave through a coupled-source tsunami event.

3.1. Description of Fault Rupture Motion

The plate uplift was tracked with a Photron Fastcam SAH 64Gb recording at 500 fps. Simultaneously, the actuator feedback was recorded. This feedback is the analog output from the encoder that reads the actuator motor response. Acquiring the video data allowed a comparison to the actuator's feedback displacement data, providing confidence in the actuator values, which have been used in the subsequent analysis [32].

The uplift motion performed by the actuator in the current investigation aims to reproduce the half-sine displacement time history of Hammack [30] with a maximum displacement of 60 mm (Figure 6a). The theoretical displacement, ζ is

$$\zeta = \frac{A}{2}\left(\sin(\omega t - \frac{\pi}{2}) + 1\right) \tag{1}$$

where A is the maximum amplitude of the uplift and $\omega = 2\pi/T$, where T is the sine wave period. Figure 6a shows a reasonable fit between the experimental data and the theoretical curve. The maximum amplitude and general shape of both curves are similar, but the measured data lags the theoretical curve.

The actuator uplift performance was not affected by the presence of the uplift plate or landslide models when performing the different test scenarios (Figure 6b): Horizontal Uplift (HU), Inclined Uplift (IU) and Inclined Uplift with Moving Landslide (IU-ML).

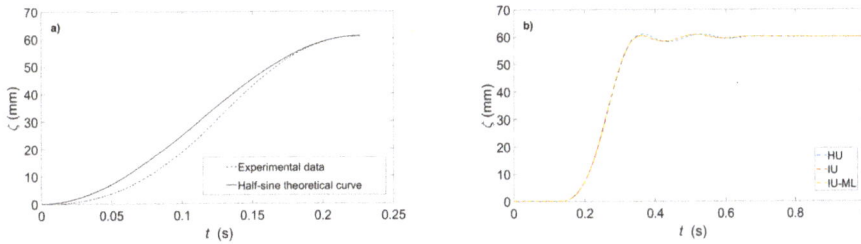

Figure 6. (a) Plate displacement (ζ) time history, comparing experimental data with a fitted half-sine curve as in Hammack [30]. (b) Plate displacement (ζ) time history for the different test scenarios: Horizontal Uplift (HU), Inclined Uplift (IU) and Inclined Uplift with Moving Landslide (IU-ML).

In contrast to Hammack [30], this study opted for an inclined uplift in order to incorporate a sliding landslide in later tests. Various incline-only (IU) combinations were tested and compared to Hammack's results, as shown later in Section 3.3. To do so, Hammack's non-dimensional parameters were adopted: amplitude scale η_0/ζ_0, where η_0 is the maximum surface elevation measured at WG2, ζ_0 is the maximum uplift distance; displacement scale ζ_0/h; size scale b/h, where where b is the fault rupture length; and the time-size ratio $t_c\sqrt{gh}/b$, where t_c is the characteristic time of motion as defined by [30] for half-sine motion. Table 2 summarises those parameters.

Table 2. Inclined uplift-only (IU) non-dimensional parameters.

η_0/ζ_0	ζ_0/h	b/h	$t_c\sqrt{gh}/b$
0.77	0.08	1.2	6.63
0.67	0.16	1.2	6.75
0.62	0.24	1.2	10.16
0.71	0.06	1.0	7.69
0.62	0.13	1.0	9.30
0.59	0.20	1.0	7.38
0.43	0.19	0.94	0.09
0.58	0.03	0.81	2.54
0.64	0.16	0.81	2.54

Hammack [30] defined three characteristic motions according to the time-size ratio: impulsive $(t_c\sqrt{gh}/b << 1)$, transitional and creeping motion $(t_c\sqrt{gh}/b >> 1)$. In the present study, the motions are predominantly transitional, with the exception of one $(t_c\sqrt{gh}/b = 0.09)$, where an uplift of 10 mm was performed in 370 mm water depth.

3.2. Description of Solid Block Motion

The landslide motion was tracked with a Nikon Digital Camera D5200 recording at 50 fps. Owing to the limited observational area (see Figure 2), the camera was placed at an oblique angle towards the landslide. Meticulous geometric camera calibration was required to obtain the camera intrinsic, extrinsic, and distortion coefficients, which were computed using the *cameraCalibrator* Matlab toolbox. These were later applied to the recorded video frames to correct the lens distortion and convert the camera pixels to real world units to obtain the correct dimensions of the tracked object. It was more feasible to apply image correction to this orientation compared with the uplift motion orientation, due to more manageable file sizes (i.e., 20.1 Mb for 8 s of video as opposed to the high-speed camera file which were around 34 Mb for 2 s of video).

The observed landslide motion for cases with no preceding uplift, is in good agreement with the following theoretical approximation (Figure 7a). A theoretical approximation for the landslide position was also obtained by considering a force balance (Appendix A), resulting in the following expression:

$$x(t) = \frac{m}{D}\ln\left(\cosh\left(\frac{\sqrt{D}\sqrt{C}}{m}t\right)\right) \tag{2}$$

where $x(t)$ is the landslide position along the slope at each instant in time t, m is the landslide mass, $D = \frac{1}{2}b^*w\rho C_d$, b^* is the landslide length, w is the landslide width, C_d is the drag coefficient, $C = W - B = mg\sin(\theta) - \rho V_l g\sin(\theta)$, θ is the slope angle, and V_l is the landslide volume.

Grilli et al. [33] also developed a theoretical approximation for the motion of a submarine landslide sliding down an incline (Equation (3)). In this case, the landslide position at each instant of time was approximated using the observed acceleration and velocity values as follows

$$S(t) = S_0\left(\ln\left(\cosh(t/t_0)\right)\right) \tag{3}$$

where $S(t)$ is the landslide position along the slope at each instant in time t, S_0 and t_0 are the characteristic position and time of the landslide during its motion, as defined by [33]. Figure 7 shows a comparison of the observed data in this study, against the theoretical approximation from [33] and the theoretical approximation developed within this study. The parameters used for the theoretical approximations are summarised in Table 3.

Table 3. Landslide models parameters for cases including a preceding uplift of 60 mm in 370 mm water depth.

Model	Mass (kg)	v_0 (m/s)	S_0 (m)	t_0 (s)	γ	C_m	C_{df}	Reynolds Number	Froude Number
ML2	8.7	1.09	0.40	0.36	3.22	0.7	0.15	3.23	0.64
ML2+	10.6	1.35	0.45	0.34	3.93	0.7	0.15	4.04	0.78
ML3	20.5	1.26	0.53	0.42	4.22	0.75	0.15	3.79	0.74
ML3B	20.5	1.16	0.45	0.39	4.22	0.75	0.15	3.49	0.68

Table 3 presents the landslide Froude number as a function of the relative density, where $Fr = v_0/\sqrt{gh}$ is the landslide Froude number, v_0 is the maximum velocity of the landslide, in relation to the relative density $\gamma = \rho_l/\rho$, where ρ_l is the landslide density and $\rho = 1000$ kg/m^3 is the density of the water. For the landslide models ML2 and ML2+, increasing the relative density results in an increase in Froude number, and therefore increasing maximum velocity. ML3 has a greater relative density than ML2+, but a smaller Froude number, which means a slower maximum velocity. This smaller velocity of ML3 might be due to an increase in friction force, which depends on the landslide weight.

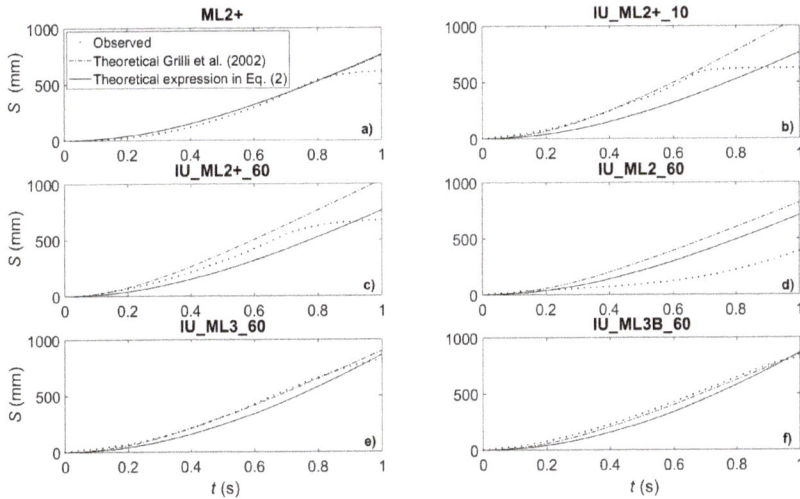

Figure 7. Observed data compared to theoretical approximations for landslide models (a) without and, (b) with an uplift of 10 mm, and (c–f) with an uplift of 60 mm in 370 mm water depth.

For all cases shown in Figure 7, the landslide motion was tracked from the same position on the slope, in 370 mm water depth. The following observations are made:

- Both theoretical approximations of landslide position show good agreement with the observed data for the landslide-only test (Figure 7a), until the point at which the landslide model experiences a change in geometry of the slope, not modelled by the theory.
- Whilst the ML2+ models (Figure 7a–c) stopped before transitioning onto the horizontal floor, the lighter model ML2 (Figure 7d) exhibited aquaplaning motion, travelling beyond the end of the slope. This resulted in longer travel times. The larger and heavier landslides ML3 and ML3B (Figure 7e,f), also exhibited longer travel times as they travelled beyond the end of the slope, due in part to their increased momentum and possibly due to deformation of the models (see final point).
- The effect of increasing uplift size on the ML2+ landslide position is also evident from these tests. The smallest uplift of 10 mm (Figure 7b), has measurements that are in very good agreement with

the theoretical approximation of Grilli et al. [33], but rather poor agreement with the theoretical approximation from the present study. However, for a 60 mm uplift, observations from the same landslide model now have worse agreement with Grilli et al. [33], though match the present theory more closely than for the smaller uplift. It is worth noting that neither of the theoretical approximations took into account the uplift motion in their formulation.

- The landslide weight was seen to have a large effect on the level of agreement of the measured and theoretical predictions. For the same uplift magnitude of 60 mm, the lightest model ML2 (Figure 7c) exhibits the worst agreement, the medium mass model ML2+ (Figure 7b) exhibits slightly better agreement, and the largest model ML3 (Figure 7e) shows by far the best agreement. N.B. Model ML3 was thicker as well as heavier.
- As described in Section 2.3, ML3 consisted of twenty slices which were joined together with tape, which flexed as the landslide hit the end of the slope, enabling it to slightly adapt its shape as it moved onto the horizontal bed. For the case of the granular landslide ML3B (Figure 7f), the landslide is seen to travel slightly more slowly than ML3 (Figure 7e), possibly due to the spreading effect of the individual slices of ML3B during its motion, as opposed to ML3, which remained as a whole.

3.3. Effect of the Uplift Displacement on the Generated Wave

Figure 8 presents the surface elevation results for an inclined uplift-only test. The generated wave is crest-led with an amplitude slightly smaller than the performed uplift (nearly 30%) at WG1, decaying in amplitude as it propagates away from the test rig backwall (Figure 2). Soon after, the leading wave reaches WG3 on the transition from the incline into the horizontal false floor, there is a slight increase in the crest amplitude, which then decreases as it propagates further. Following the crest-led wave, there is a smaller elevation trough, which appears fairly constant for different uplift displacements (same water depth), as presented later in Figure 9. As opposed to the crest, the trough increases in amplitude as it propagates away from the backwall, though soon after it reached WG3, its amplitude starts decreasing. The dependence of the incline uplift displacements relative to the water depth (ζ/h) and the generated wave amplitude in the near-field is presented in Figure 9.

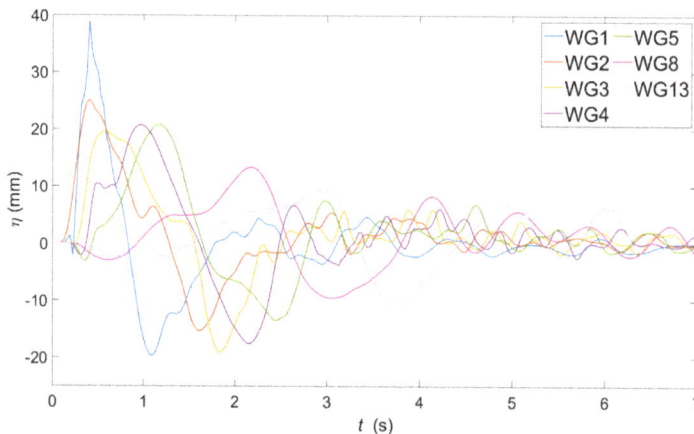

Figure 8. Surface elevation η time histories for an inclined uplift test of 60 mm in 370 mm water depth.

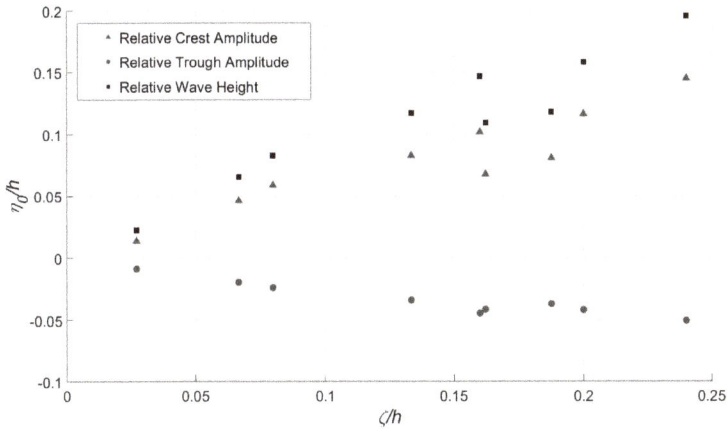

Figure 9. Relative crest and trough amplitudes (η_0/h) at WG2 as a function of displacement scales for IU only. Note that the squares represent the relative wave height.

In general, for the same uplift displacement, an increase in water depth results in a decrease of crest and trough amplitude. The effect of the uplift is more pronounced on the crest amplitude than on the trough amplitude. The transfer of the uplift displacement to crest amplitude at WG2 is 70% for 20 mm uplift, 62% for 40 mm and 58% for 60 mm, in 300 mm water depth. This means that, for the same water depth, smaller uplift displacements generate a crest amplitude almost 10% larger than for larger displacements. For the case of a 60 mm uplift in 370 mm water depth, the surface elevation was measured at WG1, showing a displacement transfer of 70%, which is similar to the measured at WG2 for a 20 mm uplift. As presented in Hammack [11], Figure 10 shows the wave profiles for an inclined-only test.

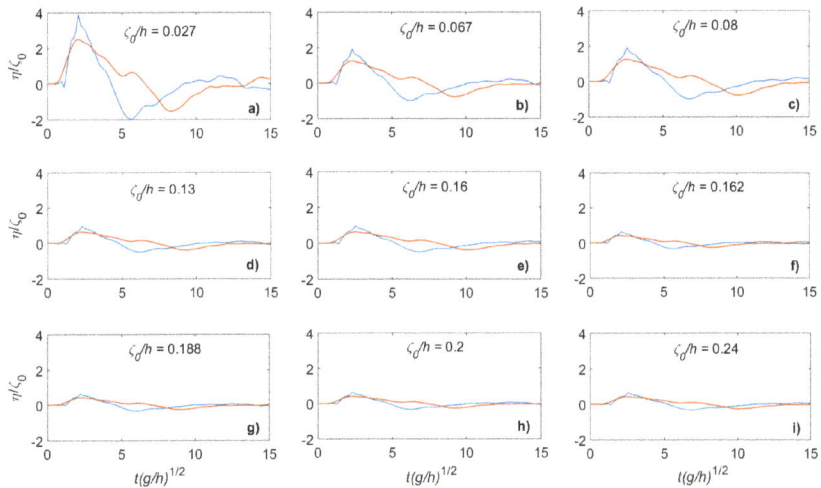

Figure 10. Dimensionless surface elevations as a function of dimensionless time at WG1 (blue) and WG2 (red) for an inclined uplift-only (IU).

13

As shown in Figure 10, the water rises to a maximum surface elevation, returning shortly after to a still water level. The maximum surface elevation reached depends upon the displacement size-ratio ζ_0/h: smaller displacement size-ratios incurred larger amplitude scales η_0/ζ_0 (Figure 10). As mentioned in Section 3.1, the uplift displacements performed in this study are predominantly transitional, with one exception. This exception is the case of a 10 mm uplift in 370 mm water depth, which with a time-size ratio significantly below 1, classified by Hammack [30] as impulsive. This case also generated the largest amplitude scale, as shown in Figure 10h.

In the present study, the most similar size-scale to Hammack's is $b/h = 1.2$. The variation of the relative amplitudes η_0/ζ_0 with the time size ratio $t_c\sqrt{gh}/b$ for a range of uplift size ratios ζ_0/h is compared to Hammack's $b/h = 1.22$ in Figure 11. Similarly to Hammack's half-sine motion, for the same time-size ratio and size-scale $b/h = 1.2$, the current study generated similar crest amplitudes, which also agree with Hammack's Linear Theory [30]. For smaller size-scales b/h, the present study generated smaller relative crest-amplitudes compared to Hammack [30], which might be due to the greater water depths tested in this study.

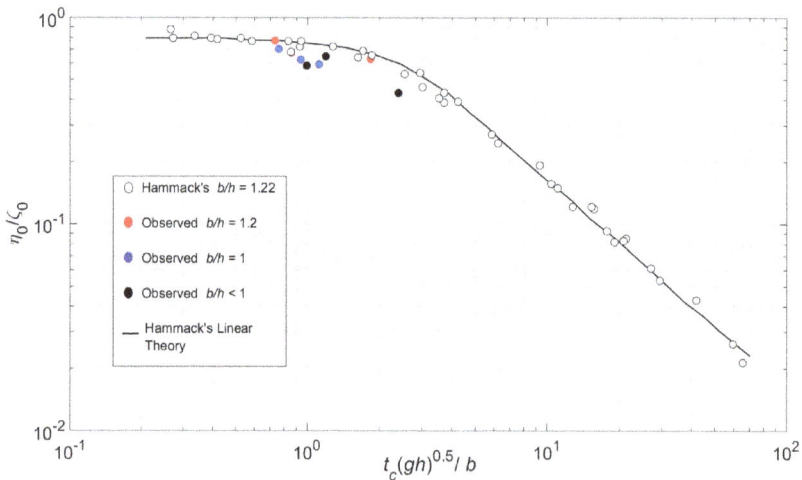

Figure 11. Variation of relative wave amplitude η_0/ζ_0 with the time size ratio $t_c\sqrt{gh}/b$ at WG2 ($x^*/h = b/h$) for a half-sine motion (IU only).

3.4. Effect of the Landslide Motion and Geometry on the Generated Wave

Figure 12 presents the surface elevation results for a landslide-only test. The generated wave is trough-led at WG1. As the trough propagates away from the back wall, its amplitude increases until it reaches WG4, close to where the landslide comes to rest, and the trough amplitude then starts decreasing. The trough is followed by a crest of smaller amplitude which propagates with a fairly constant amplitude until it reaches WG4, where its amplitude increased by almost half the initial crest amplitude. From WG6, a constant wave celerity of 1.6 m/s was observed, which was computed using the WG locations from Table 1. According the linear theory for shallow water waves (where the wavelength is much larger than the water depth), the phase velocity can be calculated as $c_P = \sqrt{gh}$, where in this case $h = 0.37$ m. With this, the theoretical phase velocity for the case presented in Figure 12 is 1.9 m/s. The difference between observed and theoretical wave celerity may be explained by dispersion effects, and/or friction induced by the flume walls.

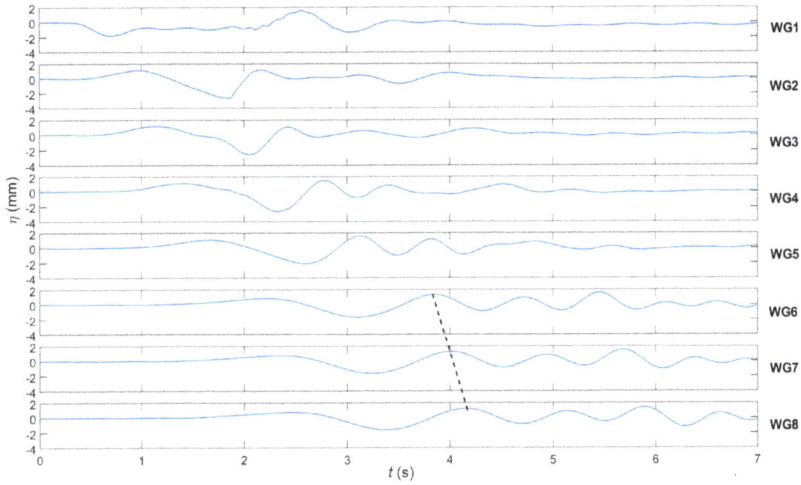

Figure 12. Surface elevation η time histories for an landslide only test in 370 mm water depth. The dashed line represents the constant wave celerity.

As mentioned in Section 2.3, three different landslide models were tested. The effect of the landslide relative density on the amplitude of the generated wave in the near-field (WG1) is illustrated in Figure 13. Increasing relative densities result in increasing crest amplitude for all three landslide cases ML2, ML2+ and ML3 with a 60 mm uplift in 370 mm water depth, being more pronounced between ML2 and ML2+. For the case of the trough amplitude, it remains unchanged between ML2 and ML2+, which have the same dimensions but different masses. However, the ML3 trough amplitude increased (absolute value of what is presented in Figure 13) with respect to ML2+. This may be explained by the fact that the thickness of ML3 is about twice the thickness of ML2+ and its mass is also greater. Notice this trend is similar to the increase in crest amplitude between ML2+ and ML3.

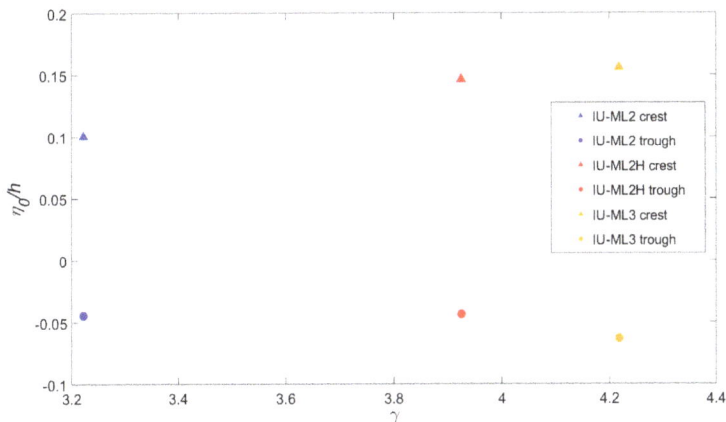

Figure 13. Normalised wave height measured at WG1 for different landslide relative density for cases with an uplift of 60 mm and 370 mm water depth.

15

For the case of the 'modular' landslide, two main configurations were tested: ML3 which is the landslide as a whole solid block, and ML3B which split into 20 slices when released. Figure 14 presents a comparison of the surface elevation changes for two tests with inclined uplift and landslide models ML3 and ML3B. The effect of the fragmentation of the landslide on the generated wave is only visible at WG1 (blue lines), where the crest amplitude is around 15 mm larger for ML3 than for ML3B. This reduction in amplitude might be due to the spreading of the landslide ML3B as it slides down the incline, meaning a reduced thickness of the landslide.

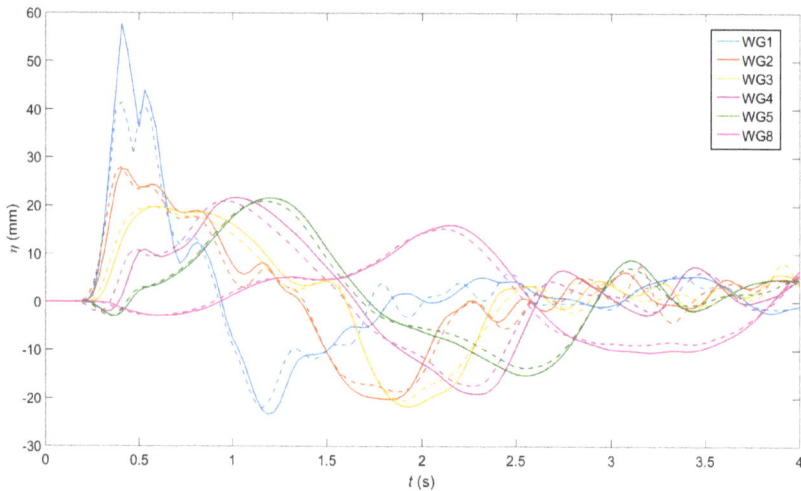

Figure 14. Surface elevation time histories for an inclined uplift of 60 mm and landslide model ML3 (solid lines), or ML3B (dashed lines) in 370 mm water depth.

3.5. Effect of a Coupled Mechanism on the Generated Wave

Figure 15 presents surface elevation profiles obtained from the wave gauges at different instants of time. Axes on the left are true scale, representing the water depth in millimetres. Axes on the right represent the surface elevation changes, which have been exaggerated ten times for clarity. Two sets of data are presented: inclined uplift-only test in grey colour and inclined uplift followed by landslide ML3 in black, both with uplift distance of 60 mm and water depth of 370 mm. The first three plots in Figure 15 capture the uplift, which happens much faster than the landslide motion, hence the small time step between them. This uplift generates the wave and contributes mainly to a crest, as mentioned earlier and illustrated in Figure 8. As soon as the landslide starts its motion, the trough of the wave increases in amplitude (from $t = 1.5$ s). Notice that the landslide front edge is always following the wave crest.

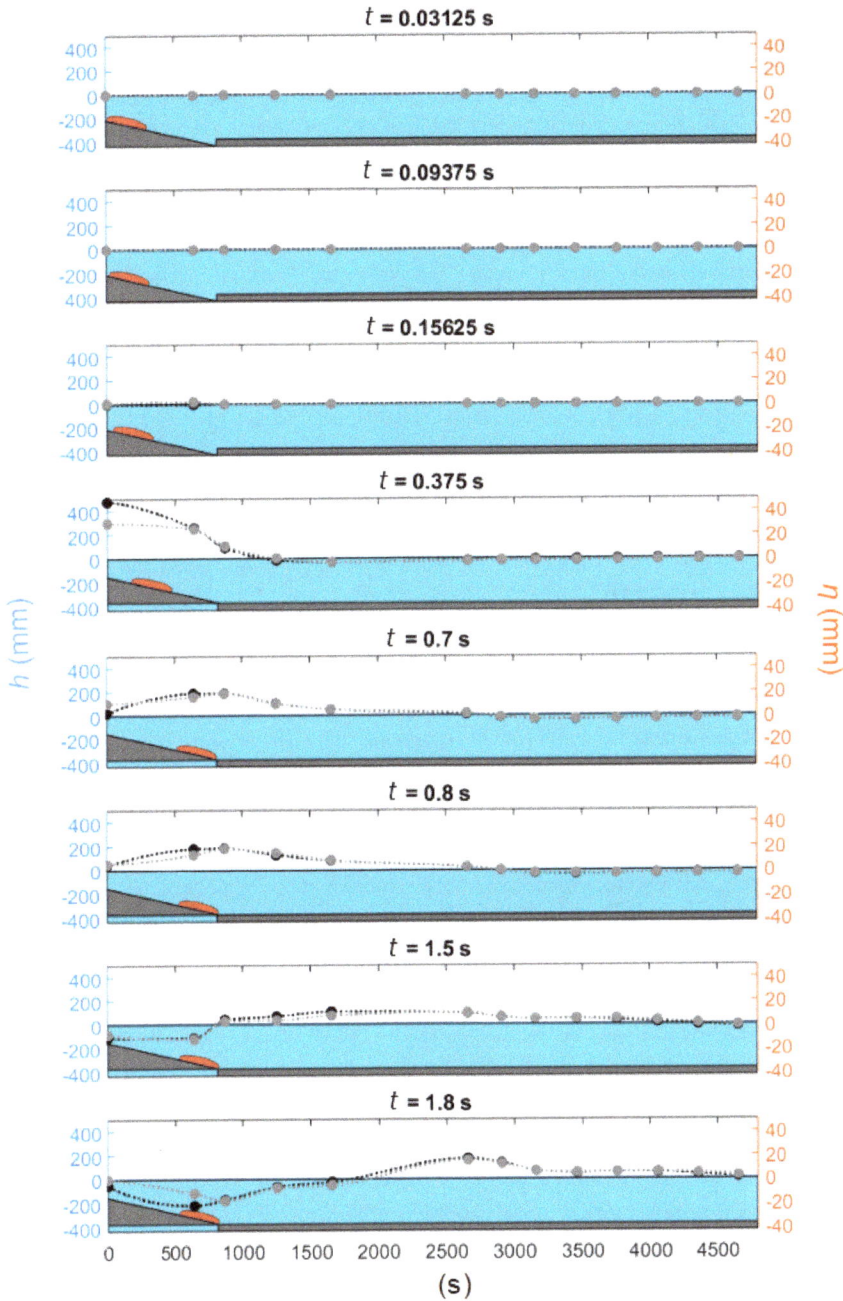

Figure 15. Stacked plots showing the setup with the surface elevation time histories for uplift-only tests (grey) and uplift followed by landslide ML3 for a 60 mm uplift and 370 mm water depth. The displacement time histories for the fault rupture are shown as the plate lifts up between $t = 0.03125$ s and $t = 0.375$ s.

4. Discussion

This paper has presented a novel physical model of a coupled-mechanism for tsunami generation involving a fault rupture, which was designed based on the prominent study by Hammack [30], and a submarine landslide, which was designed following the benchmark recommendations by Watts et al. [31]. Results from the first iteration of experiments were reported, where the coupled-sources were co-located and the landslide was released instantaneously after the fault rupture had reached the maximum uplift distance. Each source motion and the resulting wave characteristics of the source motion and generated waves have been used to understand the wave generation properties.

The current study found that for the same uplift displacement, the influence of an increasing water depth on the generated wave was a decrease of the crest amplitude, as in Hammack [30] and Todorovska & Trifunac [9]. However, owing to the space underneath the uplifted plate of this study, the generated wave amplitude was not equal to the uplift displacement. This behaviour was observed in Hammack [30], where the uplift mechanism consisted of a sealed unit, with no water exchange between chambers. It is also worth noting that the transition between the inclined and the horizontal plane seemed to have an influence on the generated wave, as explained in Section 3.3 and illustrated in Figure 8, where the generated crest amplitude decreased as it propagated away from the back wall, until it reached WG4 (transition from slope to horizontal flume floor at WG3), where an increase in amplitude was observed, to then decrease again. A similar behaviour was observed for the trough amplitude, in which the amplitude fluctuation happened at WG3.

Regarding the case of the submarine landslide only tests (without uplift), the generated wave in the near-field (WG1 and WG2) was trough led, which was also found by others such as [24,31]. In general, increasing the relative landslide density and thickness meant an increase in the crest amplitude and a slight decrease in the trough amplitude. The thickness of the landslide model was seen to have an effect on the trough amplitude of the generated wave. For the case of ML3, whose thickness was twice that of ML2, the increase in relative density meant an increase in the trough amplitude. For the case of the granular landslide, only the crest amplitude was affected by the fragmentation of the block, being around 15 mm smaller than the equivalent crest amplitude generated by ML3. As suggested in Section 3.4, the reason might have been that landslide ML2 was observed to aquaplane at the end of its motion down the incline, when it transitioned onto the horizontal plane (false floor).

The most obvious finding to emerge from the coupled-mechanism analysis is that the crest amplitude is mainly controlled by the uplift displacement, whereas the trough amplitude is influenced by the landslide motion.

Limitations and Future Work

The present study encountered various limitations, mainly due to the restricted capabilities of the test rig. The main limitation for the fault rupture model was the distance underneath the uplifted plate, under which water was recirculated. This issue was reflected in the resulting crest amplitude, particularly in inclined uplift only tests (IU), which is around 70% of the largest uplift displacement performed in the present study. In contrast, Hammack [30] performed the uplift by pushing the plate from underneath the flume floor, transferring the complete uplift distance onto the generated wave amplitude, since no volume of water was transferred elsewhere except to the generated wave. For the case of the landslide, owing to hardware limitations such as the difficulty of cable routing an accelerometer around the complex test rig, it could only be tracked using a digital camera, as opposed to placing an accelerometer inside it as in previous experimental studies dealing with submarine landslides [24,34]. There is no motion data for landslide only tests for model ML3, owing to the difficulty of releasing this model without performing the uplift, since it was extremely heavy for the retention mechanism.

Future research should be undertaken to overcome the experimental constraints. Firstly, it is to minimise or eliminate the gap underneath the fault rupture model so that the uplifted volume of water is fully transferred into the generated wave. Secondly, to overcome landslide motion tracking issues, either by having a clear observation window from which its motion can be fully tracked, or by installing a waterproof accelerometer inside the models. With these improvements, a more robust dataset could be developed and a better understanding of the generated wave properties could be obtained. Moreover, the coupled-source set up could be implemented to introduce a delay between the earthquake and the landslide, which would be closer to real cases. Furthermore, comparisons with numerical models such as SPH (Smooth Particle Hydrodynamics) would be facilitated if more robust data is provided to validate the models. A first SPH study investigating the fault rupture triggering mechanism for tsunami generation was developed by [35], showing good agreement for preliminary tests performed with an horizontal uplift only. Further simulations will investigate the inclined uplift and the inclusion of the submarine landslide.

5. Conclusions

The present study was carried out to develop a two dimensional novel coupled-source tsunami generator and to determine the generated wave properties in relation to the sources parameters. Experiments were undertaken to investigate the main feature of each source separately, which were in agreement with previous studies. These experiments confirmed that there is a noticeable influence from the secondary source (landslide) in the generation process. The key finding is that the generated wave is crest led, followed by a trough of smaller amplitude and decreasing steadily as it propagates along the flume. The crest amplitude was shown to be controlled by the fault rupture displacement scale, for which smaller displacement scales produced larger amplitude scales, whereas the trough was controlled by the landslide's relative density, where larger relative amplitudes produced larger relative trough amplitudes.

This paper has developed a novel tsunami mechanism which provided a deeper insight into the source parameters influencing a coupled-source tsunami generation scenario. These results add to the rapidly expanding field of submarine landslides as tsunami triggers, specially when combined with a co-seismic event. The major limitation of this study is the impossibility of achieving a scalable model to generate a tsunami-like wave, owing to the facility restrictions, e.g., flume tank dimensions, limited water depth, actuator placement. It was not possible to test the granular landslide without performing an uplift; therefore, it is unknown if the lack of an uplift would have influenced the landslide's motion, which could have a major effect on the generated wave. The limitations of scaling landslide-generated wave in a laboratory facility were examined by Heller et al. [36]. The scale effects were found to reduce the relative wave amplitude following the actuator placement.

Further experimental work will involve studying other spatial/dynamic features of various uplift motions, e.g., exponential, impulsive, as well as more complex landslide models, e.g., highly dispersive granular landslides. Beyond tsunami-oriented experiments, the current results will be used to develop numerical models which could facilitate the investigation of more realistic tsunami scenarios (e.g., to scale).

Author Contributions: Conceptualization, C.W.; methodology, N.P.d.P.P.; data analysis, N.P.d.P.P.; investigation, N.P.d.P.P.; writing—original draft preparation and editing, N.P.d.P.P.; writing—review, N.P.d.P.P., A.R., C.W., S.J.B.; supervision, A.R., C.W., S.J.B.; project administration, A.R.; funding acquisition, A.R.

Funding: This research was undertaken with the support of a PhD studentship jointly funded by the Marine Institute and the Sustainable Earth Institute of the University of Plymouth.

Acknowledgments: We are grateful for the technical support of the COAST Laboratory (University of Plymouth) technicians, who provided insight and expertise that greatly assisted the experiments.

Conflicts of Interest: The authors declare no conflict of interest.

Appendix A. Theoretical Formulation for Landslide Motion

Figure A1 shows a free body diagram for the sliding motion of the landslide down the incline. Considering Newton's second law $\sum F = ma$, where m is the landslide mass and a is the landslide acceleration, and the X-axis on the slope (Figure A1), then the acceleration of the landslide can be expressed as $a = \frac{d^2x}{dt^2}$.

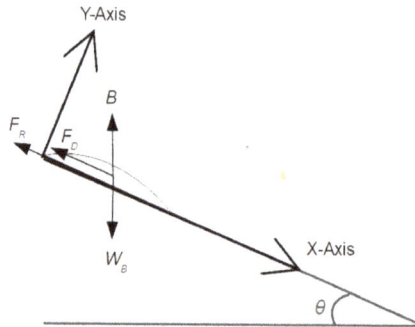

Figure A1. Free body diagram for landslide motion.

The force balance is

$$W\sin\theta - B\sin\theta - F_D - F_r = m\frac{d^2x}{dt^2} \tag{A1}$$

where W is the landslide weight, B is the buoyancy force, F_D is the drag force, F_R is the frictional force, θ is the slope angle. Re-arranging Equation (A1) gives:

$$m\frac{d^2x}{dt^2} + F_D = W\sin\theta - B\sin\theta - F_r. \tag{A2}$$

The drag force depends on the landslide velocity v to the second power $F_D = \frac{1}{2}\rho v^2 b^* w C_{df}$, where ρ is the density of water, b^* is the landslide length and w is the landslide width. To facilitate further operations, the term representing the drag force is simplified to $F_D = Dv^2$, where D represents the coefficients multiplying the velocity in the drag force expression, and $v = \frac{dx}{dt}$. The terms on the right hand side of Equation (A2) are constant and can be grouped as $C = W\sin\theta - B\sin\theta - F_r$. Grilli et al. [33] neglected the friction force to obtain his equation of motion for the submarine landslide. In the present study, as the materials employed to construct the slope have a very low friction coefficient (PTFE), the friction force is also neglected. Therefore, the term $C = W\sin\theta - B\sin\theta$, and the re-arranged second order differential equation is of the form:

$$m\frac{d^2x}{dt^2} + D\left(\frac{dx}{dt}\right)^2 - C = 0. \tag{A3}$$

Using the substitution $\frac{dx}{dt} = z$ then gives:

$$mz' + Dz^2 = C. \tag{A4}$$

Note that the apostrophe indicates derivative. This is $z' = \frac{dz}{dt}$. Equation (A4) may be rewritten as

$$z' = \frac{dz}{dt} = \frac{C}{m} - \frac{Dz^2}{m}. \tag{A5}$$

Equation (A5) can be solved by the method *'separation of variables'*.

$$\frac{dz}{\frac{C}{m} - \frac{Dz^2}{m}} = dt. \tag{A6}$$

By integrating Equation (A6), if $C, D > 0$, a particular solution can be obtained. The solution to Equation (A6) is:

$$\frac{m}{2\sqrt{C}\sqrt{D}} \ln\left(\frac{Dz - \sqrt{D}\sqrt{C}}{Dz + \sqrt{D}\sqrt{C}}\right) = t + constant. \tag{A7}$$

For $t = 0$, $z = \frac{dx}{dt} = 0$, and therefore the *constant* $= 0$. Now z is isolated as follows:

$$\ln\left(\frac{Dz - \sqrt{D}\sqrt{C}}{Dz + \sqrt{D}\sqrt{C}}\right) = t\frac{2\sqrt{C}\sqrt{D}}{m}. \tag{A8}$$

With the logarithm property $e^y = x$, $\ln(x) = y$, Equation (A8) may be rewritten as:

$$\frac{Dz - \sqrt{D}\sqrt{C}}{Dz + \sqrt{D}\sqrt{C}} = e^{t\frac{2\sqrt{C}\sqrt{D}}{m}}. \tag{A9}$$

Re-arranging the expression in Equation (A9) we obtain:

$$z = \frac{\sqrt{D}\sqrt{C}}{D} \frac{e^{\frac{2\sqrt{D}\sqrt{C}}{m}} + 1}{e^{\frac{2\sqrt{D}\sqrt{C}}{m}} - 1}. \tag{A10}$$

Note than $\tanh(x) = \frac{e^{2x}+1}{e^{2x}-1}$, and therefore Equation (A10) may be re-arranged as:

$$z = \frac{dx}{dt} = \frac{\sqrt{D}\sqrt{C}}{D} \tanh\left(\frac{\sqrt{D}\sqrt{C}}{m}t\right). \tag{A11}$$

Note that Equation (A11) represents the landslide velocity at each instant in time. The integration of Equation (A11) gives the landslide position at each instant of time:

$$x(t) = \frac{m}{D} \ln\left(\cosh\left(\frac{\sqrt{D}\sqrt{C}}{m}t\right)\right) + constant. \tag{A12}$$

Note that $\int \tanh(ax)dx = \frac{1}{a}\ln|\cosh(ax)| + constant|$. Furthermore, $x(0) = 0$, therefore $\cosh(0) = 1$ and $\ln(1) = 0$, which makes *constant* $= 0$.

References

1. Okal, E.A. Seismic parameters controlling far-field tsunami amplitudes: A review. *Nat. Hazards* **1988**, *1*, 67–96. [CrossRef]
2. Kajiura, K. Tsunami source, energy and directivity of wave radiation. *Bull. Earthq. Res. Inst.* **1970**, *48*, 835–869.
3. Fine, I.V.; Rabinovich, A.B.; Bornhold, B.D.; Thomson, R.E.; Kulikov, E.A. The Grand Banks landslide-generated tsunami of November 18, 1929: Preliminary analysis and numerical modeling. *Mar. Geol.* **2005**, *215*, 45–57. [CrossRef]
4. Kawata, Y.; Benson, B.C.; Borrero, J.C.; Borrero, J.L.; Davies, H.L.; Lange, W.P.; Imamura, F.; Letz, H.; Nott, J.; Synolakis, C.E. Tsunami in Papua New Guinea was as intense as first thought. *Eos Trans. Am. Geophys. Union* **1999**, *80*, 101–105. [CrossRef]
5. Tappin, D.R.; Watts, P.; McMurtry, G.M.; Matsumoto, T. Offshore evidence on the source of the 1998 Papua New Guinea tsunami: A sediment slump. In Proceedings of the International Tsunami Symposium, Seattle, WA, USA, 7–10 August 2001; Volume 175, pp. 381–388.

6. Synolakis, C.E.; Bardet, J.P.; Borrero, J.C.; Davies, H.L.; Okal, E.A.; Silver, E.A.; Sweet, S.; Tappin, D.R. The slump origin of the 1998 Papua New Guinea Tsunami. *Proc. R. Soc. Math. Phys. Eng. Sci.* **2002**, *458*, 763–789. [CrossRef]

7. Heinrich, P.; Guibourg, S.; Roche, R. Numerical modeling of the 1960 Chilean tsunami. Impact on French Polynesia. *Phys. Chem. Earth* **1996**, *21*, 19–25. [CrossRef]

8. Stein, R.S. The role of stress transfer in earthquake occurrence. *Nature* **1999**, *402*, 605–609. [CrossRef]

9. Todorovska, M.I.; Trifunac, M.D. Generation of tsunamis by a slowly spreading uplift of the sea floor. *Soil Dyn. Earthq. Eng.* **2001**, *21*, 151–167. [CrossRef]

10. Kowalik, Z.; Knight, W.; Logan, T.; Whitmore, P. The tsunami of 26 December, 2004: Numerical modeling and energy considerations. *Pure Appl. Geophys.* **2007**, *164*, 379–393. [CrossRef]

11. Hammack, J.L. A note on tsunamis: Their generation and propagation in an ocean of uniform depth. *J. Fluid Mech.* **1973**, *60*, 769–799. [CrossRef]

12. Iwasaki, S. Experimental study of a tsunami generated by a horizontal motion of a sloping bottom. *Bull. Earthq. Res. Inst.* **1982**, *57*, 239–262.

13. Jamin, T.; Gordillo, L.; Ruiz-Chavarria, G.; Berhanu, M.; Falcon, E. Experiments on generation of surface waves by an underwater moving bottom. *Proc. R. Soc. Math. Phys. Eng. Sci.* **2015**, *471*. [CrossRef]

14. Heller, V.; Hager, W.H. Impulse product parameter in landslide generated impulse waves. *J. Waterw. Port Coast. Ocean. Eng.* **2010**, *136*, 145–155. [CrossRef]

15. Heller, V.; Spinneken, J. Improved landslide-tsunami prediction: Effects of block model parameters and slide model. *J. Geophys. Res. Ocean.* **2013**, *118*, 1489–1507. [CrossRef]

16. Heller, V.; Bruggemann, M.; Spinneken, J.; Rogers, B.D. Composite modelling of subaerial landslide–tsunamis in different water body geometries and novel insight into slide and wave kinematics. *Coast. Eng.* **2016**, *109*, 20–41. [CrossRef]

17. Evers, F.M.; Hager, W. Generation and spatial propagation of landslide generated impulse waves. *Coast. Eng. Proc.* **2017**, *1*, 13. [CrossRef]

18. Assier-Rzadkieaicz, S.; Heinrich, P.; Sabatier, P.C.; Savoye, B.; Bourillet, J.F. Numerical modelling of a landslide-generated tsunami: The 1979 Nice event. *Pure Appl. Geophys.* **2000**, *157*, 1707–1727. [CrossRef]

19. Tinti, S.; Pagnoni, G.; Zaniboni, F. The landslides and tsunamis of the 30th of December 2002 in Stromboli analysed through numerical simulations. *Bull. Volcanol.* **2006**, *68*, 462–479. [CrossRef]

20. Heinrich, P.; Piatanesi, A.; Hébert, H. Numerical modelling of tsunami generation and propagation from submarine slumps: The 1998 Papua New Guinea event. *Geophys. J. Int.* **2001**, *145*, 97–111. [CrossRef]

21. Lynett, P.; Liu, P.L.F. A numerical study of submarine-landslide-generated waves and run-up. *Proc. R. Soc. Math. Phys. Eng. Sci.* **2002**, *458*, 2885–2910. [CrossRef]

22. Bondevik, S.; Løvholt, F.; Harbitz, C.; Mangerud, J.; Dawson, A.; Svendsen, J.I. The Storegga Slide tsunami—Comparing field observations with numerical simulations. In *Ormen Lange—An Integrated Study for Safe Field Development in the Storegga Submarine Area*; Elsevier: New York, NY, USA, 2005; pp. 195–208.

23. Watts, P. Tsunami features of solid block underwater landslides. *J. Waterw. Port Coast. Ocean. Eng.* **2000**, *2*, 144–152. [CrossRef]

24. Sue, L.P. Modelling of Tsunami Generated by Submarine Landslides. Ph.D. Thesis, University of Canterbury, Christchurch, New Zealand, 2007.

25. Van Nieuwkoop, J.C.C. Experimental and Numerical Modelling of Tsunami Waves Generated by Landslides. Master's Thesis, Delft University of Technology, Delft, The Netherlands, 2007.

26. Wiegel, R.L. Laboratory studies of gravity waves generated by the movement of a submerged body. *Eos Trans. Am. Geophys. Union* **1955**, *36*, 759–774. [CrossRef]

27. Watts, P.; Grilli, S.T. Underwater landslide shape, motion, deformation, and tsunami generation. *Int. Offshore Polar Eng. Conf.* **2003**, *5*, 364–371.

28. Heller, V.; Hager, W.H. Wave types of landslide generated impulse waves. *Ocean. Eng.* **2011**, *38*, 630–640. [CrossRef]

29. Mohammed, F.; Fritz, H.M. Physical modeling of tsunamis generated by three-dimensional deformable granular landslides. *J. Geophys. Res. Ocean.* **2012**, *118*, 3221. [CrossRef]

30. Hammack, J.L. Tsunamis—A Model of Their Generation and Propagation. Ph.D. Thesis, California Institute of Technology, Pasadena, CA, USA, 1972.

31. Philip, W.; Fumihiko, I.; Aaron, B.; Grilli, S.T. Benchmark cases for tsunamis generated by underwater landslides. In Proceedings of the Fourth International Symposium on Ocean Wave Measurement and Analysis, San Francisco, CA, USA, 2–6 September 2001. [CrossRef]
32. Perez del Postigo, N.; Raby, A.; Whittaker, C.; Boulton, S.J. Tsunami generation by combined fault rupture and landsliding. In Proceedings of the 7th International Conference on the Application of Physical Modelling in Coastal and Port Engineering and Science (Coastlab18), Santander, Spain, 22–26 May 2018; pp. 1–10.
33. Grilli, S.T.; Vogelmann, S.; Watts, P. Development of a 3D numerical wave tank for modeling tsunami generation by underwater landslides. *Eng. Anal. Bound. Elem.* **2002**, *26*, 301–313. [CrossRef]
34. Whittaker, C.; Nokes, R.; Davidson, M. Tsunami forcing by a low Froude number landslide. *Environ. Fluid Mech.* **2015**, *15*, 1215–1239. [CrossRef]
35. Wana, R.; Perez, N.; Hughes, J.; Graham, D.; Raby, A.; Whittaker, C. Smoothed Particle Hydrodynamics (SPH) modelling of tsunami waves generated by a fault rupture. In Proceedings of the ASME 2007 26th International Conference on Offshore Mechanics and Arctic Engineering, San Diego, CA, USA, 10–15 June 2019.
36. Heller, V.; Hager, W.H.; Minor, H.E. Scale effects in subaerial landslide generated impulse waves. *Exp. Fluids* **2008**, *44*, 691–703. [CrossRef]

© 2019 by the authors. Licensee MDPI, Basel, Switzerland. This article is an open access article distributed under the terms and conditions of the Creative Commons Attribution (CC BY) license (http://creativecommons.org/licenses/by/4.0/).

Journal of
Marine Science and Engineering

MDPI

Article

A Numerical Landslide-Tsunami Hazard Assessment Technique Applied on Hypothetical Scenarios at Es Vedrà, Offshore Ibiza

Hai Tan [1,2], Gioele Ruffini [1], Valentin Heller [1,*] and Shenghong Chen [2]

[1]   Environmental Fluid Mechanics and Geoprocesses Research Group, Faculty of Engineering,
     University of Nottingham, Nottingham NG7 2RD, UK; Hai.Tan@whu.edu.cn (H.T.);
     Gioele.Ruffini1@nottingham.ac.uk (G.R.)
[2]   State Key Laboratory of Water Resources and Hydropower Engineering Science, Wuhan University,
     Wuhan 430072, China; chensh@whu.edu.cn
*   Correspondence: Valentin.Heller@nottingham.ac.uk; Tel.: +44(0)-115-748-6049

Received: 15 August 2018; Accepted: 25 September 2018; Published: 28 September 2018

Abstract: This study presents a numerical landslide-tsunami hazard assessment technique for applications in reservoirs, lakes, fjords, and the sea. This technique is illustrated with hypothetical scenarios at Es Vedrà, offshore Ibiza, although currently no evidence suggests that this island may become unstable. The two selected scenarios include two particularly vulnerable locations, namely: (i) Cala d'Hort on Ibiza (3 km away from Es Vedrà) and (ii) Marina de Formentera (23 km away from Es Vedrà). The violent wave generation process is modelled with the meshless Lagrangian method smoothed particle hydrodynamics. Further offshore, the simulations are continued with the less computational expensive code SWASH (Simulating WAves till SHore), which is based on the non-hydrostatic non-linear shallow water equations that are capable of considering bottom friction and frequency dispersion. The up to 133-m high tsunamis decay relatively fast with distance from Es Vedrà; the wave height 5 m offshore Cala d'Hort is 14.2 m, reaching a maximum run-up height of over 21.5 m, whilst the offshore wave height (2.7 m) and maximum inundation depth at Marina de Formentera (1.2 m) are significantly smaller. This study illustrates that landslide-tsunami hazard assessment can nowadays readily be conducted under consideration of site-specific details such as the bathymetry and topography, and intends to support future investigations of real landslide-tsunami cases.

Keywords: Es Vedrà; landslides; landslide-tsunamis; numerical modelling; smoothed particle hydrodynamics; SWASH; wave propagation

---

1. Introduction

Landslides are a main source of impulse waves in reservoirs [1,2], lakes [3,4], fjords [5], and the sea [6]. Such impulse waves are better known as landslide tsunamis if they occur in the open sea, and for simplicity, this term is adopted herein irrespectively of the type of water body in question. Landslide tsunamis involve several complex processes such as slide propagation and impact into a water body, tsunami generation, propagation, and run-up, which often results in casualties and destroyed infrastructure. Landslide tsunamis from nearshore to offshore involve multiple time and length scales such that more than one numerical code is required in order to reliably deal with all of these scales, e.g., DualSPHysics for the nearshore [7,8] combined with SWASH (Simulating WAves till SHore) [9,10] for the offshore region.

Landslide tsunamis occurred for example in Lake Askja, on Iceland, on 21 July 2014. An approximately 20 million m^3 large landslide initiated a 50-m large tsunami resulting in a run-up height of 60–80 m on the opposite shore [4]. Another example is a 270 million m^3 landslide impacting into the Vajont reservoir in Italy, on 9 October 1963. The resulting wave overtopped the dam crest, resulting in approximately 2000 casualties [1]. Such cases demonstrate the need for reliable landslide-tsunami hazard assessment methods to prevent similar catastrophes from the large number of potential landslides in proximity of a water body all around the world [11,12].

Landslide tsunamis are mainly investigated by means of (i) physical modelling and (ii) numerical simulations. (i) Generic physical experiments are conducted either in wave flumes (two-dimensional, 2D) or wave basins (three-dimensional, 3D), by using either blocks or granular materials to represent the slide [13]. Empirical equations are then derived to predict the generated waves based on the slide characteristics (impact velocity, volume, thickness, slope angle) and water depth [14–16]. For example, the empirical equations of both the maximum wave height and the wave run-up derived from experiments summarised by Heller et al. [17] were successfully applied to a number of real scenarios by Battaglia et al. [18], BGC Engineering Inc. [19], Fuchs et al. [2], Gabl et al. [20], Lüthi and Vieli [21], and Oppikofer et al. [22]. Generic experiments rely on a number of simplifications such as an idealised slide mass, bathymetry, and geometry. Whilst these simplifications are often sufficient for initial estimates, they are sometimes insufficient for reliable predictions in more complex cases [2] where prototype specific physical case studies may be a better, but significantly more time-consuming and expensive option [17].

(ii) In numerical modelling, on the other hand, the effect of the bathymetry and topography can readily be taken into account. The solid phase (landslide) is often represented by a Newtonian [12] or non-Newtonian fluid [23], with the more realistic discrete element method (DEM) becoming increasingly popular [24,25]. The fluid phase (tsunamis) is either described by mesh-based or particle-based methods [24,26,27].

In the traditional Euler mesh-based method, the mesh is fixed and requires an additional surface tracking algorithm to reconstruct the free surface. In contrast, for Lagrangian mesh-based methods, such as the particle finite element method (PFEM) [23,28], the mesh moves with the physical material. In PFEM, re-meshing is required due to the severe mesh distortion that occurs when the material experiences large deformation and displacement [23,28].

With regard to the particle-based method, moving particle semi-implicit (MPS) [26] and smoothed particle hydrodynamics (SPH) [7,24,27] are widely used. SPH is a Lagrangian meshless method that is well suited for modelling violent free surface flows such as landslide tsunamis. However, in comparison to mesh-based methods, SPH involves many particles, resulting in large computational costs. This often restricts its application to two dimensions, which is suboptimal for landslide tsunamis, as it neglects the important lateral and spatial wave propagation [16,17,29].

Recently, DualSPHysics (the successor of SPHysics) [30], a hardware accelerated open source code running both with central processing units (CPUs) and graphics processing units (GPUs), has been developed to overcome this problem. Crespo et al. [30] found that the GPU version of the code is up to two orders of magnitudes faster than the serial CPU version, e.g., a GTX Titan GPU card may speed up the process by up to a factor of 149 compared to a single core Xeon X5500 CPU. This enables the investigation of real-world engineering problems such as landslide tsunamis [7,8].

Whilst SPH commonly deals well with violent free surface flows, it is less well-suited for long-distance water wave propagation [31]. SPH is thus often coupled with a wave propagation model, which also reduces the computational cost significantly. Boussinesq-type models such as FUNWAVE(-TVD) (TVD version of the fully nonlinear Boussinesq wave model) are particularly well-suited to model landslide-tsunami propagation given that these waves are highly dispersive [32]. Non-hydrostatic non-linear shallow water equations (NLSWEs) models, such as SWASH, are able to model dispersive waves as well, namely via the non-hydrostatic pressure distribution and the addition of a vertical momentum equation to the shallow water equations. Non-hydrostatic NLSWEs models usually need a relatively large number of horizontal layers to resolve the frequency dispersion up to an acceptable level [9]. However, SWASH also uses in addition a compact difference scheme allowing for an accurate modelling of frequency dispersion, even with two layers [10]. SWASH is thus a good alternative to the commonly recommended Boussinesq-type models for landslide-tsunami propagation.

Examples relying on a coupling approach include Narayanaswamy et al. [33], who proposed a hybrid SPHysics-FUNWAVE model, and Altomare et al. [34], who proposed a hybrid DualSPHysics-SWASH model to study wave propagation from offshore to nearshore. Viroulet et al. [35] used SPHysics for landslide-tsunami generation, and Gerris, which is a two-phase finite volume approach, for tsunami propagation. However, these simulations [33–35] were conducted in two dimensions. Three-dimensional (3D) simulations, as presented by Abadie et al. [12], are much rarer. They combined a Navier–Stokes model (THETIS) and FUNWAVE-TVD to study the 3D tsunami generation and propagation from the potential collapse of the Cumbre Vieja Volcano, La Palma, on the Canary Islands.

In this work, a 3D landslide-tsunami hazard assessment technique is illustrated by taking advantage of the two distinct models DualSPHysics (for slide impact and wave generation) and SWASH (for wave propagation). The feasibility of the proposed coupling method is demonstrated by investigating hypothetical landslide-tsunami scenarios originating from Es Vedrà, offshore Ibiza, although currently no evidence suggests that this island may become unstable. Es Vedrà has been selected because of its steep flanks exceeding typical basal friction angles >30° and for its proximity to other islands, allowing for run-up and inundation investigations. That landslide tsunamis at Es Vedrà have not been investigated previously added also to the motivation to select this particular case. More importantly, the developed techniques in this work are transferable to other cases in reservoirs, lakes, fjords, and the sea. Figure 1 shows the plan view of Ibiza and Formentera with Es Vedrà, including the paths of two critical landslide-tsunami scenarios investigated herein, and a picture of Es Vedrà taken from Ibiza. These two scenarios were selected because inhabitants, tourists, and infrastructure at these low-lying locations are particularly exposed to potential tsunami run-up in contrast to more elevated areas elsewhere on Ibiza and Formentera. These two scenarios involve tsunamis propagating towards (i) Cala d'Hort (3 km away from Es Vedrà) and (ii) Marina de Formentera (23 km away from Es Vedrà).

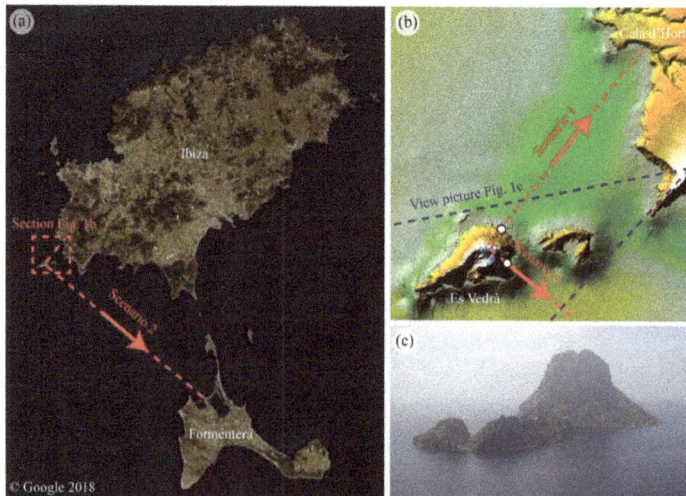

Figure 1. (a) Plan view of Ibiza and Formentera with Es Vedrà and tsunami propagation towards Cala d'Hort (scenario 1) and Marina de Formentera (scenario 2) marked with red dashed lines (adapted from Google maps); the red dashed square highlights the region shown in (**b**). (**b**) Reproduced bathymetry [36] and topography [37] in proximity of Es Vedrà with the range covered by the picture in (**c**) marked with blue dashed lines. (**c**) Picture of Es Vedrà taken by V. Heller from Ibiza.

The remaining sections of this article are organised as follows. Section 2 presents the basic features of DualSPHysics and SWASH. The results of the convergence tests, as well as the wave generation and propagation in the two scenarios, are described in Section 3. In Section 4, the results are compared with another study, the implications of this work on landslide-tsunami hazard assessment are discussed, and the limitations are highlighted. The main conclusions are finally presented in Section 5.

2. Methods

The slide movement and landslide-tsunami generation were conducted with DualSPHysics v4.0, University of Vigo, weakly coupled with SWASH v4.01, Delft University of Technology, which was used for the wave propagation. The numerical backgrounds of these two open source codes are introduced in Sections 2.1 and 2.2.

2.1. DualSPHysics

2.1.1. Basic Principles

The open source code DualSPHysics v4.0 was used herein [30], which is based on weakly compressible SPH (WCSPH). The fluid phase in DualSPHysics is governed by the Navier–Stokes equations with the partial differential equations reduced to ordinary differential equations in a Lagrangian framework. The conservation of mass and momentum are expressed in Equations (1) and (2) [31,38]:

$$\frac{d\rho}{dt} = -\rho \nabla \cdot v \tag{1}$$

$$\frac{dv}{dt} = -\frac{1}{\rho}\nabla p + g + \mathit{\Gamma} \tag{2}$$

where ρ is the density, v is the velocity vector, p is the pressure, g is the gravitational acceleration vector, and $\mathit{\Gamma}$ is the dissipative term. DualSPHysics provides two options for dissipative terms, namely: artificial viscosity and laminar viscosity with sub-particle scale turbulence [30]. These two viscosity treatments provided similar results for the dam break case [39]. Thus, the artificial viscosity is widely used in landslide-tsunami modelling, resulting in the reasonable agreement of the numerical wave profiles with experimental results [8,40]. Based on this, the artificial viscosity was applied in the present work.

In SPH, the fluid is discretised into a set of particles carrying properties such as density, pressure, velocity, position, etc. In general, two steps are required to transform Equations (1) and (2) into the SPH formalism, i.e., a kernel approximation and a particle approximation [41]. In the first step, any field variable f associated with particle a (located at x_a) can be represented by an integral at location x in the form of Equation (3):

$$f(x_a) = \int_{\Omega} f(x)W(x_a - x, h_p)dx \tag{3}$$

In Equation (3), Ω is the computation domain, W is the weighting function or smoothing kernel, which monotonically decreases with distance, and h_p is the smoothing length determining the size of the kernel support. The kernels cubic spline and Wendland are available in DualSPHysics, and in this work, the former has been used.

In the second step, the integral in Equation (3) is approximated by interpolating the characteristics of the surrounding particles resulting in Equation (4):

$$f(x_a) \approx \sum_b f(x_b)\frac{m_b}{\rho_b}W_{ab} \tag{4}$$

where the summation is over all the particles in the kernel support. W_{ab} is short for $W(x_a - x_b, h_p)$, and m_b and ρ_b are the mass and density, respectively, of particle b (located at x_b).

Similarly, the derivative of the field variable f can be expressed as shown in Equation (5):

$$\nabla f(x_a) \approx \sum_b f(x_b) \frac{m_b}{\rho_b} \nabla_a W_{ab} \tag{5}$$

where ∇_a is the derivative with respect to the coordinates of particle a.

2.1.2. Governing Equations

The governing equations in the SPH formalism are briefly discussed with details given by Crespo et al. [30]. The WCSPH method, although it is able to resolve the fluid kinematics well, suffers from unphysical pressure oscillations. In WCSPH, the pressure is linked to the density by means of an equation of state. Hence, remedies are proposed to stabilise the density field, and thus ensuring that the pressure field is noise-free. This is achieved with a density filter [38] and a density diffusion correction [42]. In the density diffusion correction, which is also referred to as the delta SPH algorithm, a diffusion term is added into the continuity equation to eliminate the numerical noise on the pressure field.

The continuity equation including the density diffusion correction is written in Equation (6):

$$\frac{d\rho_a}{dt} = \sum_b m_b v_{ab} \cdot \nabla_a W_{ab} + 2\delta h_p \sum_b \frac{m_b}{\rho_b} \bar{c}_{ab} (\rho_a - \rho_b) \frac{1}{x_{ab}^2 + 0.01h_p^2} x_{ab} \cdot \nabla_a W_{ab} \tag{6}$$

where subscript a and b denote fluid particles a and b, respectively, ρ_a is the density of the fluid particle a, $v_{ab} = v_a - v_b$, v_a and v_b are the velocity vectors of fluid particles a and b, respectively, δ is the delta SPH coefficient, $\bar{c}_{ab} = 0.5(c_a + c_b)$, c_a and c_b are the speed of sound at the locations of fluid particles a and b, respectively, $x_{ab} = x_a - x_b$, x_a and x_b are the position vector of fluid particles a and b, respectively, and $0.01h_p^2$ is added to prevent singularities. The second term at the right-hand side of Equation (6) is the density diffusion term.

The momentum equation is written in the form of Equation (7) [30,38]:

$$\frac{dv_a}{dt} = -\sum_b m_b \left(\frac{p_a}{\rho_a^2} + \frac{p_b}{\rho_b^2} + \Pi_{ab} \right) \nabla_a W_{ab} + g \tag{7}$$

where p_a and p_b are the pressure of fluid particles a and b, respectively, and Π_{ab} is the artificial viscosity, which accounts for the effects of dissipation, and is given by Equation (8):

$$\Pi_{ab} = \begin{cases} \frac{-\kappa \bar{c}_{ab} \mu_{ab}}{\bar{\rho}_{ab}} & v_{ab} \cdot x_{ab} < 0 \\ 0 & v_{ab} \cdot x_{ab} > 0 \end{cases} \tag{8}$$

In Equation (8), κ is the artificial viscosity coefficient, $\mu_{ab} = h_p v_{ab} \cdot x_{ab} / \left(x_{ab}^2 + 0.01h_p^2 \right)$ and $\bar{\rho}_{ab} = 0.5(\rho_a + \rho_b)$.

In contrast to incompressible SPH where the pressure is solved by the Poisson equation, the pressure in WCSPH is determined via an equation of state. This Equation (9) relates the pressure to the density of the fluid to close the governing equation system [30,38]:

$$p_a = \frac{c_0^2 \rho_0}{\gamma} \left[\left(\frac{\rho_a}{\rho_0} \right)^\gamma - 1 \right] \tag{9}$$

In Equation (9), γ is typically selected as 7, $\rho_0 = 1000$ kg/m^3 is the reference density, and $c_0 = .c(\rho_0) = \sqrt{\partial p_a / \partial \rho_a}|_{\rho_0}$ is the speed of sound at the reference density. The relative density fluctuation can be related to c_0 and the maximum velocity of the SPH particles v_{max}, namely $|\Delta \rho|/\rho \sim v_{max}^2 / c_0^2$ [43]. Therefore, c_0 should be set to be at least 10 times larger than the maximum flow velocity such that the relative density fluctuation can be constrained to less than 1%, satisfying the condition that the fluid is nearly incompressible.

Equations (6), (7), and (9) are complete, and result in the density, velocity, and pressure of each fluid particle by using an integration scheme. In DualSPHysics, the two integration schemes Verlet and Symplectic are provided [30], whilst this work adopted the Verlet scheme.

Each fluid particle position is then updated according to its velocity as shown in Equation (10):

$$\frac{dx_a}{dt} = v_a - \varepsilon \sum_b \frac{m_b}{\rho_{ab}} v_{ab} W_{ab} \tag{10}$$

where $\varepsilon = 0.5$. The second term on the right-hand side of Equation (10) is the XSPH variant [44], which is aimed at moving each fluid particle with a velocity close to the average of its neighbourhood.

The representation of solid boundaries in SPH is challenging primarily due to the truncation of the kernel support near a boundary. In general, solid boundaries are represented by particles, and three kinds of boundary particles are widely used in the technical literature, namely: (i) repulsive, (ii) ghost, and (iii) dynamic particles.

In terms of (i), repulsive boundary conditions, a repulsion force, normally in the Lennard–Jones form, is felt by those fluid particles in proximity of the boundary to prevent its motion beyond the domain of interest [45]. However, the repulsive particles have no contribution to the density of the fluid particles, which may further contaminate the pressure field. (ii) Ghost particles, which can either be dynamically generated or predefined, are placed beyond the boundary to fill the truncated domain of the kernel [46]. The field variables of the ghost particles are then mirrored from the inner fluid particles. Method (ii) has proven to be effective for simple boundaries, but encounters difficulties in dealing with more complex boundaries [31]. In (iii), dynamic boundary conditions [47], the dynamic boundary particles are forced to satisfy the same equations as the fluid particles, but are fixed in position, providing a sufficient repelling force to the fluid particles nearby. In this work, dynamic boundary conditions were used, given that this is currently the only option in DualSPHysics [30].

The motion of the landslide is fully resolved with no initial velocity imposed. The solid phase (landslide) is assumed to be rigid [13] and treated as a floating object [30,38]. The translational and rotational motion was obtained according to the force and momentum exerted on the floating object, i.e., by boundaries or ambient fluid particles. All of the relevant parameters that are used in DualSPHysics are summarised in Table 1. This includes the particle spacing, number of particles, ratio of smoothing length to particle spacing, delta-SPH coefficient, artificial viscosity coefficient, reference density ρ_0, dimensionless parameter γ, ε, coefficient of speed of sound, number of time steps applied to Eulerian equations, the Courant–Friedrichs–Lewy (CFL) number, and physical time.

Table 1. Parameters used in DualSPHysics. The values marked with * were only used in the convergence tests. CFL: Courant–Friedrichs–Lewy, SPH: smoothed particle hydrodynamics.

Parameter	Scenario 1	Scenario 2
Particle spacing dp (mm)	10.0	7.5 */10.0/15.0 */20.0 *
Number of particles (million)	8.40	26.81 */11.74/3.52 */1.66*
Smoothing kernel (-)		Cubic spline kernel
Smoothing length/particle spacing (-)		1.732
Density correction (-)		Delta-SPH algorithm
Delta-SPH coefficient (-)		0.1
Dissipative term (-)		Artificial viscosity
Artificial viscosity coefficient (-)		0.05
ρ_0 (kg/m^3)		1000
γ (-)		7
ε (-)		0.5
Coefficient of speed of sound (-)		17
Boundary conditions (BCs) (-)		Dynamic BCs
Time integration algorithm (-)		Verlet scheme
Number of time steps applied to Eulerian equations (-)		40
CFL number (-)		0.2
Physical time (s)		3.0

The 3D numerical models of scenario 1 (Cala d'Hort) and scenario 2 (Marina de Formentera) in DualSPHysics were reconstructed in a similar manner. First, the bathymetric data available from the European Marine Observation and Data Network [36] and topographic data from the Spanish National Center of Geographic Information (CNIG) [37] were visualised and superimposed with each other. Second, the superimposed bathymetry and topography were interpolated such that the resolution was 5 m × 5 m. Third, the 4000 m × 4000 m large domain of interest was selected, and the slide plane was defined on Es Vedrà. Fourth, the slide profile and the interpolated bathymetry and topography were scaled at 1:500 before importing them into DualSPHysics. At this reduced scale, the simulations were run with the same particle spacing and similar SPH parameters as in Heller et al. [8], i.e., some experience was available. Last, the water body was added into the scaled model.

In DualSPHysics, a series of wave probes were placed at x = 200 m, 300 m, 500 m, 750 m, 1000 m, etc., to record wave profiles along the slide axis, with the coordinate origin for x placed at the intersection of the still water level (SWL) with the slide axis (white circles in Figures 1b and 2). The coupling location was selected by inspecting the numerical wave profiles, and the first location was selected where both the wave crest and trough are fully developed such that SWASH can cope with the input time series. This is the case at x = 750 m in scenario 1 and at x = 1000 m in scenario 2. Alternative coupling criteria, e.g., based on the location of the maximum wave amplitude defined in Heller and Hager [14], are discussed in Ruffini et al. [48].

One wave probe was placed on the slide axis, and 25 were placed perpendicularly on either side on a straight line. This resulted in a total of 51 wave probes separated by 30 m. The wave profiles and wave kinematics outputs from DualSPHysics were used as input in SWASH using a time series for each of the 51 grid points and depth-averaged velocities over both layers. The total length of the wave generation boundary in both scenarios was 1500 m.

2.2. SWASH

2.2.1. Numerical Background

SWASH v4.01 [9,10,49] was used in this study to simulate the wave propagation in both scenarios. SWASH is a numerical model based on the depth-averaged non-hydrostatic NLSWEs given in Equations (11)–(13):

$$\frac{\partial \eta}{\partial t} + \frac{\partial d\bar{u}}{\partial x} + \frac{\partial d\bar{v}}{\partial y} = 0, \tag{11}$$

$$\frac{\partial \bar{u}}{\partial t} + \bar{u}\frac{\partial \bar{u}}{\partial x} + \bar{v}\frac{\partial \bar{u}}{\partial y} + g\frac{\partial \eta}{\partial x} + \frac{1}{d}\int_{-h}^{\eta}\frac{\partial q}{\partial x}dz + c_f\frac{\bar{u}\sqrt{\bar{u}^2 + \bar{v}^2}}{d} = \frac{1}{d}\left(\frac{\partial d\tau_{xx}}{\partial x} + \frac{\partial d\tau_{xy}}{\partial y}\right), \tag{12}$$

$$\frac{\partial \bar{v}}{\partial t} + \bar{u}\frac{\partial \bar{v}}{\partial x} + \bar{v}\frac{\partial \bar{v}}{\partial y} + g\frac{\partial \eta}{\partial y} + \frac{1}{d}\int_{-h}^{\eta}\frac{\partial q}{\partial y}dz + c_f\frac{\bar{v}\sqrt{\bar{u}^2 + \bar{v}^2}}{d} = \frac{1}{d}\left(\frac{\partial d\tau_{yx}}{\partial x} + \frac{\partial d\tau_{yy}}{\partial y}\right), \tag{13}$$

where t is time, x, y, and z are the coordinates located at the mean SWL, g is the gravitational acceleration, $h(x,y)$ is the still water depth, $\eta(x,y,t)$ is the water surface elevation from the SWL, and $d = h + \eta$ is the total water depth. $\bar{u}$ and $\bar{v}$ are the depth-averaged flow velocities in the two main directions. $\tau_{xx} = 2v_t\partial\bar{u}/\partial x$, $\tau_{xy} = \tau_{yx} = v_t(\partial\bar{v}/\partial x + \partial\bar{u}/\partial y)$, and $\tau_{yy} = 2v_t\partial\bar{v}/\partial y$ are the horizontal turbulent stresses where $v_t(x,y,t)$ is the horizontal eddy viscosity defined in Zijlema et al. [10]. c_f is the bottom friction coefficient defined by Manning's formula, and $q(x,y,z,t)$ is the non-hydrostatic pressure. Equations (11)–(13) were expanded for the multi-layer case by Stelling and Zijlema [48] using a discretisation method based on the Keller-box scheme.

The non-hydrostatic pressure is defined as a term of the total pressure in Equation (14) [50]:

$$p_t = g(\eta - z) + q = p_h + q, \tag{14}$$

where p_h is the hydrostatic term. The hydrostatic balance is given by Equation (15):

$$\frac{\partial p_h}{\partial z} = -g. \tag{15}$$

The computation of the integral of the non-hydrostatic pressure gradient in Equations (12) and (13) is introduced in Zijlema et al. [10] where the free surface boundary condition of the non-hydrostatic pressure is $q|_{z=\eta} = 0$ and at the bottom the non-hydrostatic pressure is defined by applying the Keller-box method. Then, the vertical velocities at the free surface w_s and at the bottom w_b are considered with their respective momentum equations. Here, the vertical acceleration is instantaneously determined from the non-hydrostatic pressure. Finally, by combining the vertical momentum equations with the non-hydrostatic pressure equation at the bottom and using the kinematic bottom boundary condition $w_b = -\bar{u}\partial h/\partial x - \bar{v}\partial h/\partial y$, the conservation of local mass yields Equation (16):

$$\frac{\partial \bar{u}}{\partial x} + \frac{\partial \bar{v}}{\partial y} + \frac{w_s - w_b}{d} = 0. \tag{16}$$

Equation (16) closes the equation system and enables, together with the boundary conditions, to solve Equations (11) to (13).

2.2.2. Numerical Model Setup and Boundary Conditions

The use of SWASH v4.01 for the wave propagation allowed reducing the computational time significantly in both scenarios, and helped to avoid the wave decay problem of DualSPHysics (Section 4.2). Two different regular grids were created, one for each scenario, both with a grid resolution of $\Delta x = \Delta y = 30$ m, allowing at least 30 grid points per wavelength, which is sufficient to ensure the convergence of the solution [48,49]. The bathymetry was retrieved from the European Marine Observation and Data Network [36] with a resolution of approximatively 200 m. The topography was taken from the CNIG [37] with digital terrain model data with an accuracy of 25 m to match the grid resolution closely. The created domains counted 94 × 94 grid points for scenario 1 (Cala d'Hort) and 900 × 400 grid points for scenario 2 (Marina de Formentera). The Delft3D QUICKIN v5.0, Deltares, Delft, has been used to create the grids by processing the data with a triangular interpolation. Furthermore, smoothing of the interpolation has been applied on the west side of the Cap Llentrisca topography to avoid instabilities in the simulations from the large slopes at this location.

The code has been compiled for the use with multiple CPUs. All of the simulations were performed on the Nottingham High Performance Computing (HPC) cluster Minerva. For the biggest domain investigated, a real time of 25 min took 8.25 h of simulation time using four CPU cores and 10 GB of RAM. To solve the equations using multiple cores, the model divided the computational domain into subdomains. A stripwise decomposition method along the propagation direction of the tsunami was chosen for these purposes.

All of the simulations were performed using two layers, which generates a maximum error of 1% in the phase velocity for waves with $kd \leq 3$ [10], where k specifies the wave number. An accurate simulation of the frequency dispersion can be very important for landslide tsunamis, as they can be highly dispersive [32]. In addition, the breaking criterion with default parameters has been applied to ensure that the energy dissipation due to wave breaking is accurately simulated.

Both water surface and a depth-averaged velocity time series were used as input for each point, as described in Section 2.1. The water surface time series was assigned using a weakly reflective boundary condition [51], and the depth-averaged velocity was assigned as a time series using velocity components that were directed perpendicularly to the coupling line. To achieve this, the velocity output from DualSPHysics was decomposed, and only the required component was used. However, it is expected that this did not affect the results significantly, as the grids were oriented along the main tsunami propagation direction. Furthermore, all of the open sides of the domain were defined using a Sommerfeld boundary condition [52], which allows waves to cross the outflow boundary without reflections, and is particularly suited for long waves. This condition, for a boundary oriented parallel to the y-axis, is given in Equation (17):

$$\frac{\partial u}{\partial t} + \sqrt{g\bar{d}}\frac{\partial u}{\partial x} = 0, \tag{17}$$

The Manning's roughness coefficient n was used to include bottom friction via Equation (18):

$$c_f = \frac{n^2 g}{d^{1/3}}. \tag{18}$$

This study applies the default value $n = 0.019 \text{ s/m}^{1/3}$ as recommended for wave simulation on sandy beaches [53]. Further, a value of 5 mm is chosen as the threshold for the minimum computed water depth.

Finally, an explicit time discretisation method is used. This method uses the CFL condition to adjust the time step during the simulation accordingly. The Courant number C_r for the performed simulations is defined in Equation (19):

$$C_r = \Delta t \left(\sqrt{g\bar{d}} + \sqrt{u^2 + v^2} \right) \sqrt{\frac{1}{\Delta x^2} + \frac{1}{\Delta y^2}} \leq 1, \tag{19}$$

where Δx and Δy are the distances between two grid points in the x and y directions, respectively. To calculate whether to increase or reduce the time step, a minimum and maximum C_r threshold can be applied in the simulation in order to control the convergence of the solution accurately. $C_{r,min} = 0.1$ and $C_{r,max} = 0.5$ are used herein as suggested for large, nonlinear waves and wave interaction with structures and steep slopes [53].

3. Results

3.1. Wave Generation

3.1.1. Slide Scenarios

Figure 2 shows the two-dimensional (2D) slide profiles along the main wave propagation directions (Figure 1a). The origins are defined at the intersections of the island, the slide axis and the SWL (Figure 1b). Es Vedrà is mainly composed of Miocene evaporates [54], which is a special type of limestone. The Miocene limestone has a density of 1210 kg/m³ to 2510 kg/m³ [55]. In this work, the slide density is assumed to be 1600 kg/m³. The slide profile is obtained by cutting the island at an angle of 30°, assuming that the sliding surface is planar. Smaller and larger slope angles in 5° intervals were also investigated. Smaller slope angles result in larger slide volumes, but in smaller slide impact velocities. No slide movement was observed for very flat angles. In scenario 1, 30° resulted in the largest tsunamis. In scenario 2, both slope angles 25° and 30° resulted essentially in the same wave heights such that again, 30° was selected to be consistent with scenario 1. The slide volume in scenario 1 was 4.33×10^6 m³, corresponding to a mass of 6.93×10^9 kg and in scenario 2, the volume was 6.92×10^6 m³ and the mass 11.08×10^9 kg.

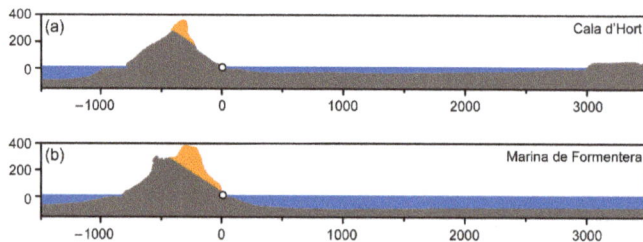

Figure 2. Central sections of the slide profiles marked in light brown for (**a**) scenario 1 and (**b**) scenario 2.

3.1.2. Convergence Tests

Convergence tests have been conducted for scenario 2. Five wave probes were placed along the slide axis at 0.40 m, 0.60 m, 1.00 m, 1.50 m, and 2.00 m. Four particle resolutions were selected with the resulting wave profiles at 2.00 m, as shown in Figure 3, thereby focusing on the primary wave. The results in Figure 3 are shown at scale 1:500, while all of the other results in this article are upscaled to the nature scale with Froude scaling [56,57]. Generally speaking, the wave profiles converge with finer particle resolution towards a larger wave amplitude, except for dp = 7.5 mm. The wave heights for dp = 7.5 mm, 10 mm, 15 mm, and 20 mm are 0.137 m, 0.142 m, 0.119 m, and 0.101 m, respectively. In relation to the wave height observed at dp = 7.5 mm, this results in differences of 3.1%, −13.3%, and −26.5% for dp = 10 mm, 15 mm, and 20 mm, respectively. The discrepancy of the wave height between dp = 7.5 mm and 10 mm is relatively small, indicating convergence. Based on this, the particle resolution 10 mm was selected as a compromise between accuracy and computational cost. This resolution was then applied to both scenarios, which resulted in approximately 8.40 million particles in scenario 1 and 11.74 million particles in scenario 2 (Table 1). The computation (excluding post-processing) took 80 min for scenario 1 and 115 min for scenario 2 on a Titan Xp GPU for a scaled real time of 3.0 s (67 s at nature scale).

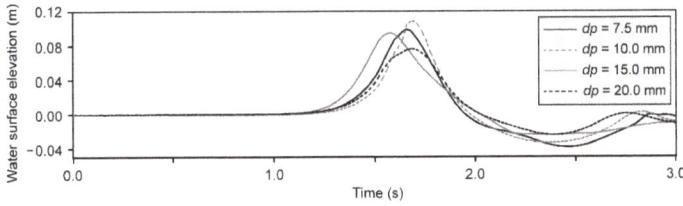

Figure 3. Convergence test: wave profiles at 2.00 m along the slide axis for different particle spacings *dp* in scenario 2.

3.1.3. Analysis of the Results

Figure 4 shows the slide kinematics in both scenarios upscaled to the nature scale. The slide front impact velocity in scenario 1 is 49.48 m/s, which is close to the maximum slide velocity. In scenario 2, the slide front impact velocity is 13.02 m/s, with the slide velocity further increasing by approximately a factor of four. It is emphasised that the slide velocity in DualSPHysics may overestimate the real slide velocity [8]. This is because the slide block was treated as a floating object, and the basal friction was not fully modelled (Section 4.2). However, the maximum slide velocities shown in Figure 4 are not unrealistic for landslides, e.g., the slide impact velocity of the 1958 Lituya Bay rockslide was estimated at 92 m/s [14], and of the 2007 Chehalis landslide was estimated at 60 m/s [58]. The velocities in Figure 4 are considered to be extreme case scenarios where the initial friction coefficients are reduced, which is a phenomenon known as hypermobility, resulting sometimes in anomalously high velocities and long runout distances in nature for slide volumes >10^6 m^3 [59]. The slide position time histories, which were directly derived from the slide velocity time histories, are also shown in Figure 4. The slide positions from initiation to deposition in both scenarios exceed 1000 m, which further indicates hypermobility features of the herein investigated landslides.

The wave propagations at the nature scale are depicted in Figure 5 (scenario 1) and Figure 6 (scenario 2). In both scenarios, the water is fully displaced at the wake of the landslide. This is an artifact of the discrete SPH particles, and is not expected to occur to such a degree in reality. In scenario 1, before t = 16.9 s, the slide moves below the primary wave crest, indicating that the slide motion is faster than the wave. This is also observed in scenario 2, where the primary wave finally overtakes the landslide at t = 20.1 s, and continues to propagate outwards (Figure 6d). Both Figures 5 and 6 confirm that the landslide tsunamis are largest on the slide axis and significantly smaller at the peripheries [29,48].

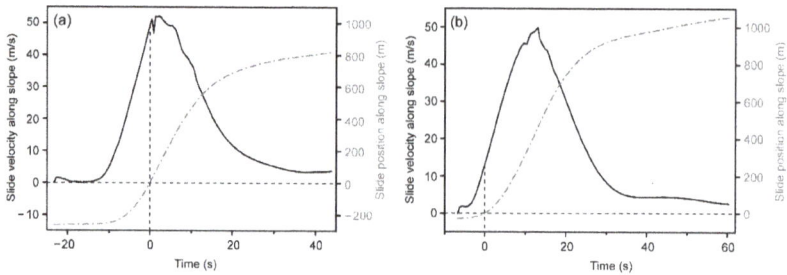

Figure 4. Time history of the slide velocity and position for (**a**) scenario 1 and (**b**) scenario 2; the landslide fronts reach the water surface at position = 0 m and $t = 0$ s.

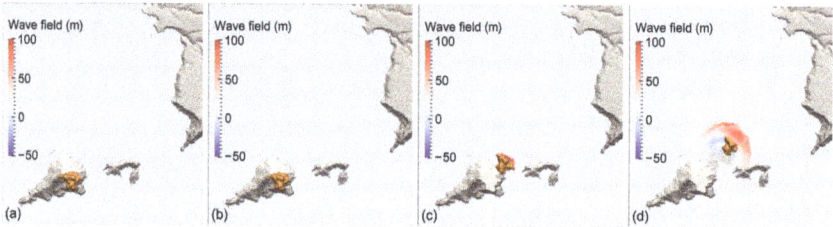

Figure 5. Time history of wave generation in scenario 1 at (**a**) $t = -23.3$ s, (**b**) -9.9 s, (**c**) 3.5 s, and (**d**) 16.9 s.

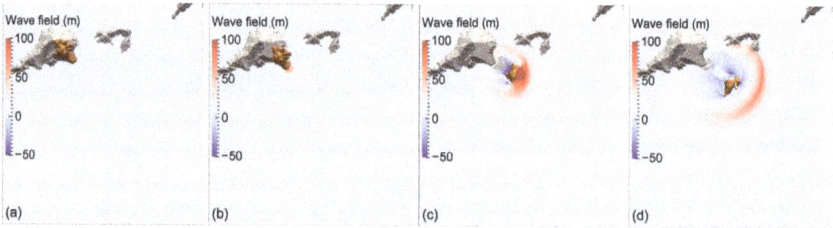

Figure 6. Time history of wave generation in scenario 2 at (**a**) $t = -6.7$ s, (**b**) 6.7 s, (**c**) 20.1 s, and (**d**) 33.5 s.

Four wave probes in scenario 1 and five wave probes in scenario 2 were placed along the slide axes (Figure 1) to record the landslide tsunamis, as shown in Figure 7. The flat parts in Figure 7 indicate that the water at these wave probe locations is fully displaced, as discussed in the last paragraph. The maximum wave height in scenario 1 of 133.0 m is larger than the 75.4 m that is observed in scenario 2 (Figure 7). However, the landslide mass is with 6.93×10^9 kg smaller in scenario 1 than in scenario 2, in which it is 11.08×10^9 kg. This seemingly contradiction can be explained with the larger slide Froude number $v/(gh)^{1/2}$ in scenario 1, which has a significantly more dominant effect on the tsunami magnitude than the slide mass for subaerial landslides, according to the impulse product parameter [14].

The maximum wave amplitude in scenario 1 was more than twice as large as the water depth (Figure 2a). This is not unusual in the slide impact zone for large slide impact velocities (see e.g., Figure 7a,b in Huang et al. [60] or Figure 5a in Heller and Hager [14], where the relative maximum wave amplitudes were similarly large). Such large "waves" are the consequences of the impact crater, water splash, and the slide trapped below the primary wave, and does not represent a stable wave. Figure 7 also reveals that the wave decays faster in scenario 1; the water column is elevated by the landslide located below the primary wave at the impact zone, and the sudden drop in wave elevation may be related to the disappearance of this effect once the wave travels faster than the slide.

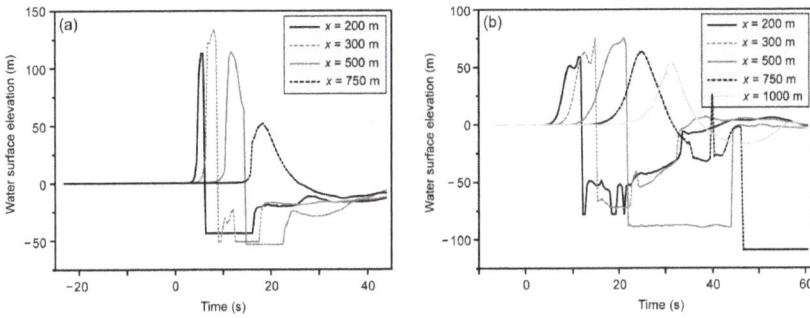

Figure 7. Time history of the water surface elevation measured along the slide axes; (**a**) scenario 1 resulting in a maximum wave amplitude of 133.0 m and (**b**) scenario 2 resulting in a maximum wave amplitude of 75.4 m.

3.2. Wave Propagation and Run-Up

3.2.1. Scenario 1 (Cala d'Hort)

Figure 8a shows the time series of the water surfaces measured every 200 m along the slide axis for 10 locations in total. The primary wave remains the largest wave up to the beach with a maximum wave amplitude of 28.5 m at $x = 850$ m. This is around 40% less than the input value that was used in SWASH at $x = 750$ m, which is consistent with the decay found between $x = 500$ m and $x = 750$ m in DualSPHysics (Figure 7a). The minimum primary wave amplitude in Figure 8a of 8.7 m is found at $x = 2650$ m. Figure 8b shows a time series at different distances from Cala d'Hort, with $x_b = 0$ m corresponding to the still water shoreline ($x = 2980$ m in the global coordinate system). The maximum wave amplitude of 14.2 m is observed at $x_b = -5$ m, with a noticeable increase from $x = 2650$ m (Figure 8a), which was mainly due to a second wave caused by reflection from the cliffs surrounding the beach (Figure 9) and shoaling effects. The effect of the reflection can be seen at $x_b = -100$ m and $t = 135$ s in Figure 8b.

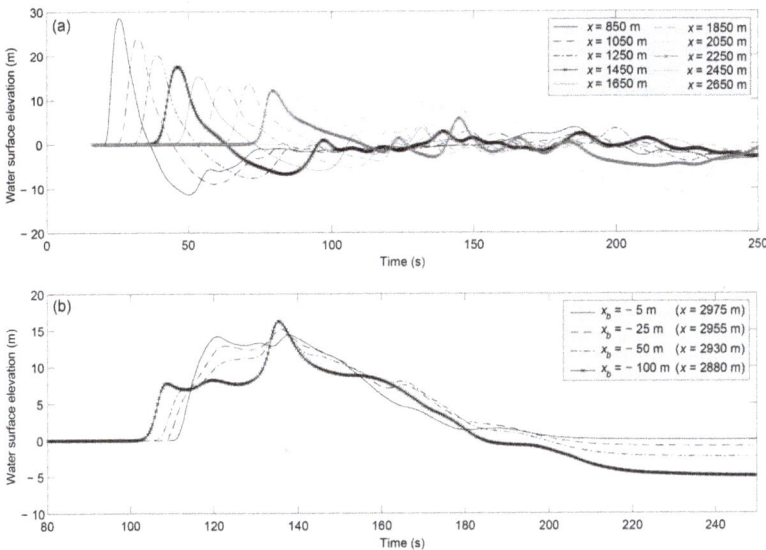

Figure 8. Time history of the water surface elevation measured along the slide axis for scenario 1; (**a**) relative to the slide impact point ($x = 0$ m) and (**b**) relative to Cala d'Hort ($x_b = 0$ m and $x = 2980$ m).

Figure 9 shows the wave propagation towards Cala d'Hort with four snapshots at different times. This shows that the tsunami front remains circular and the maximum amplitude occurs on the slide axis. Further, Figure 9d shows that the wave reaches the shore 122 s after slide impact. Figure 9c,d clearly confirm the reflections on the southeast cliffs generating the secondary peak in Figure 8b.

The tsunami run-up is investigated next with Figure 10, and five critical points are examined in more detail. The values shown refer to the wet computational grid points with P1 and P4 showing the wet most inland points. Given that the grid resolution is only 30 m, the maximum run-up height is likely to be underestimated by the specified values. P1 is positioned inland of Cala Carbo, which is a smaller beach north of Cala d'Hort where at a terrain elevation of 6 m a maximum inundation depth of 0.23 m is measured. P2 is positioned on the west side of the island Escull de Cala d'Hort where the tsunami generates a run-up height of 5.7 m with a maximum water depth of 4.7 m. The last three points in Figure 10 are all positioned at Cala d'Hort. Point B is on the slide axis near the beach shore where a maximum inundation depth of 14.43 m on a terrain elevation of 0.42 m is measured. The water level starts to decrease towards P3 at the most inland part of the beach with a run-up of 9.24 m and a maximum inundation depth of 11.5 m over the entire event. Finally, the water reaches the maximum terrain elevation of 19.7 m above SWL at P4 and a maximum inundation depth of 1.8 m.

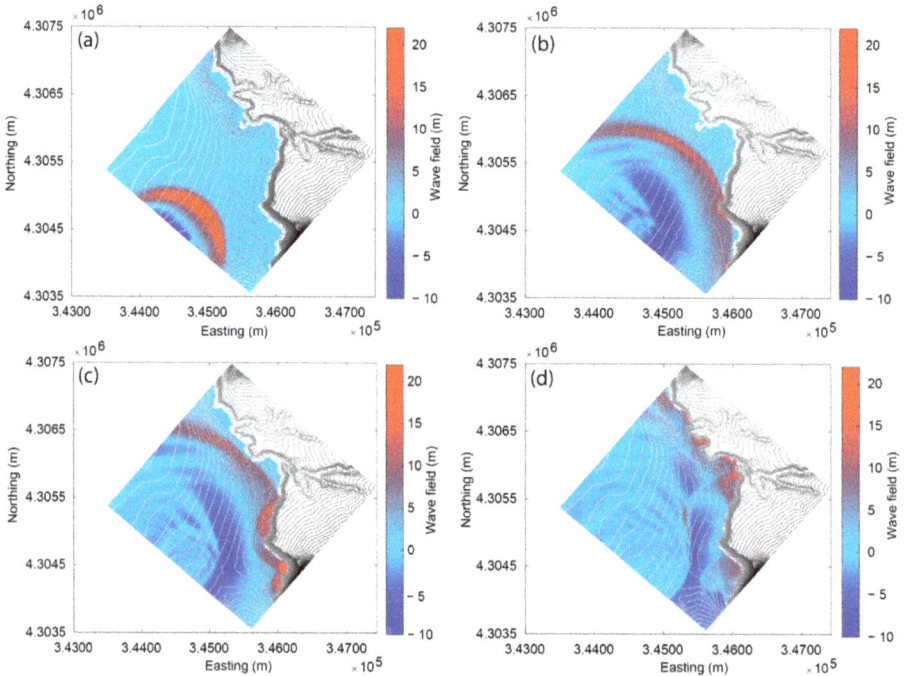

Figure 9. Time history of wave propagation in scenario 1 at (**a**) $t = 33$ s, (**b**) 66 s, (**c**) 96 s, and (**d**) 122 s.

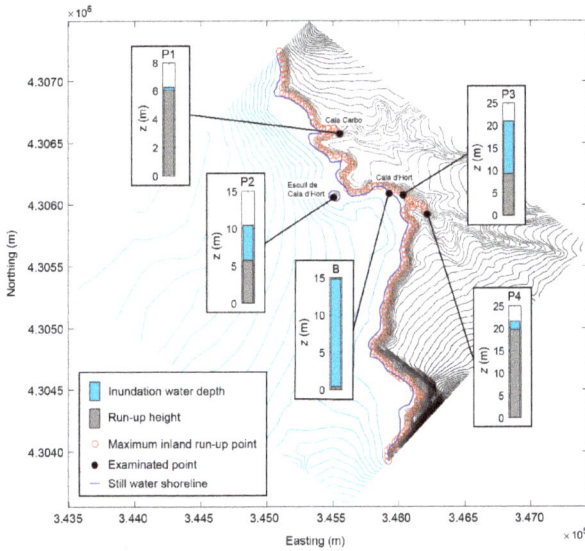

Figure 10. Maximum run-up heights and water depths at different points of interest in scenario 1.

3.2.2. Scenario 2 (Marina de Formentera)

The tsunami propagation and run-up analysis was performed similarly for scenario 2 as for scenario 1. However, the distance between the slide impact and Marina de Formentera is around 10 times larger. Therefore, Figure 11a shows the water surface elevations on the slide axis for every 2000 m only. The maximum amplitude at $x = 1500$ m is 31 m. The wave decays fast until $x = 13,500$ m; afterwards, the amplitude stabilises at approximately 2.5 to 3 m whilst approaching the harbour, as the wave decay may be compensated by shoaling effects.

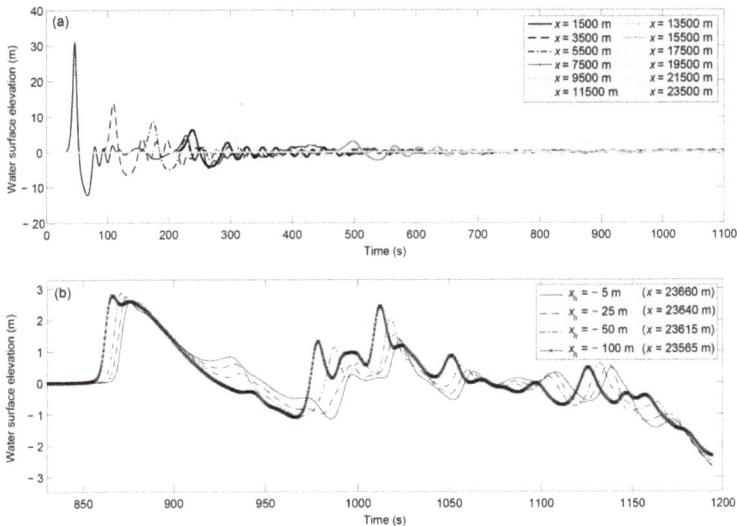

Figure 11. Time history of the water surface elevation measured along the slide axis for scenario 2; (**a**) relative to the slide impact point ($x = 0$ m) and (**b**) relative to Marina de Formentera ($x_h = 0$ m and $x = 23,665$ m).

Figure 11b shows how the tsunami approaches the tip of the harbour with a constant wave amplitude of 2.7 m due to the moderate sea bottom slope. Further, Figure 12 shows four snapshots of the tsunami propagation towards Formentera, where the primary tsunami impacts the harbour and coast of the island in Figure 12d after 15 min. It should be noted that the range of the legends in Figure 12b–d have been changed in relation to Figure 12a to reflect the strong reduction in wave amplitude. Figure 12b shows the tsunami diffraction around Cap Llentrisca on Ibiza, which subsequently results in reflection from the cliffs.

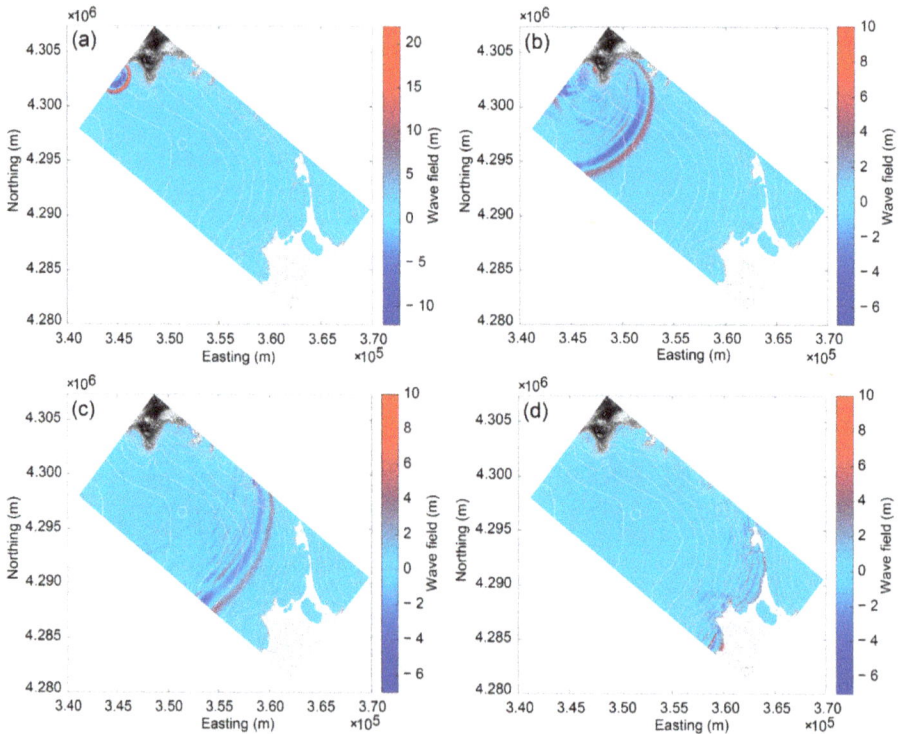

Figure 12. Time history of wave propagation in scenario 2 at (**a**) $t = 65$ s, (**b**) 305 s, (**c**) 605 s, and (**d**) 905 s.

The run-up in scenario 2 is investigated in Figure 13, showing how the tsunami floods Marina de Formentera (H), Estany Pudent (P2), and Playa de Illetas (P3), with the maximum horizontal inundation represented by red circles. The elevation at point H is 1.5 m, and the water level reaches 2.7 m above SWL. Further, noticeable run-up affects Cala Saona (P1) on the south, with a terrain elevation of 5.57 m and an inundation depth of 0.57 m. In P2, the tsunami flows into the salt lagoon Estany Pudent, where a water depth of 0.77 m above the terrain elevation can be seen. Then, the tsunami propagates in the lagoon, and dissipates further energy with travel distance as shown in P4, with a tsunami height of only 0.04 m. Finally, parts of Playa de Illetas are flooded, as highlighted with point P3 in Figure 13.

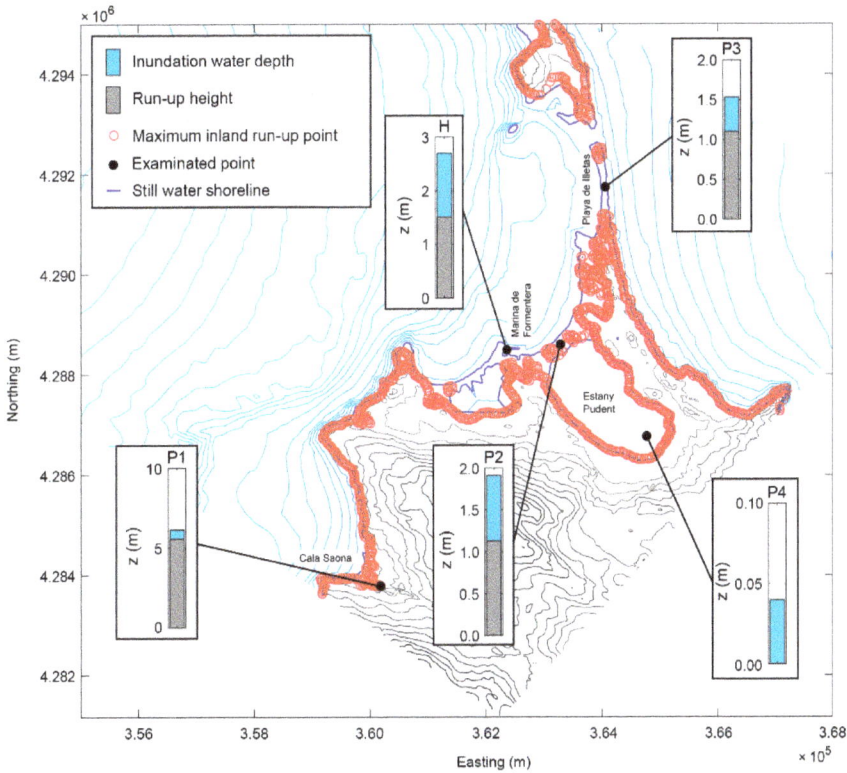

Figure 13. Maximum run-up heights and water depths at different points of interest in scenario 2.

4. Discussion

4.1. Comparison with La Palma Case

Abadie et al. [12] conducted a closely related numerical study by investigating a potential landslide tsunami from La Palma, Canary Islands, and expressed the tsunami decay as a function of the radial distance r from the slide impact location. The tsunami amplitudes a of the present study in both scenarios are compared in Figure 14 with the wave amplitude decays found in Abadie et al. [12]. The amplitudes are shown for the same positions as in Figures 7, 8a and 11a. The solid and dashed lines represent the maximum and minimum decay found by Abadie et al. [12] for slide volumes of 80 km^3 and 450 km^3, respectively. The tsunami amplitude of scenario 1 decays similarly to $r^{-1.19}$, while the tsunami in scenario 2 decays with $r^{-0.95}$, and lays close to the lower boundary found by Abadie et al. [12]. This small difference is likely due to the different bathymetries in the two cases. Scenario 2 shows a larger water depth and a very mildly sloped seabed, whereas in scenario 1, the shallower water depth and rapidly varying seabed interact more with the tsunami, increasing the decay rate. Mainland Ibiza, which is shown on the north side in Figure 12a, may also have a small effect on the lateral energy spread [29,48], and the wave decay may also change with the wave type (landslide-tsunami propagation in 3D typically involves Stokes-like and cnoidal-like waves [29]). Overall, a consistent wave amplitude decay between the present study and Abadie et al. [12] is found.

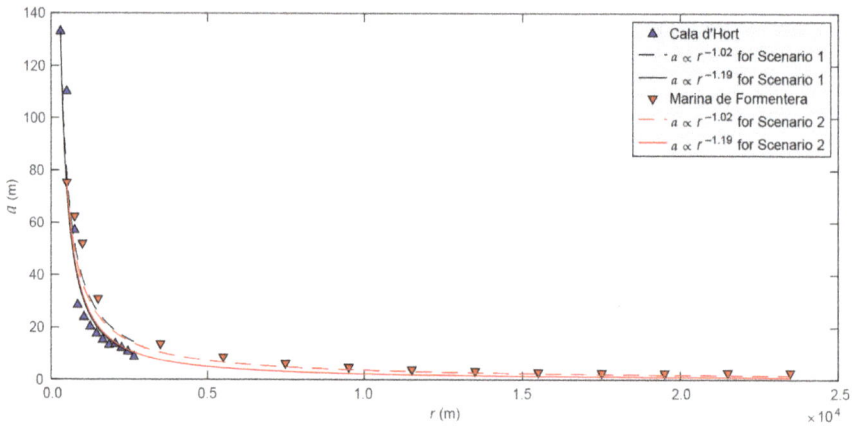

Figure 14. Comparison of wave amplitude *a* decay over radial distance *r* for both scenarios with maximum and minimum decay found by Abadie et al. [12].

4.2. Implications and Limitations

The main purpose of this work is to demonstrate a versatile numerical technique that can be applied to other landslide-tsunami events in reservoirs, lakes, fjords, and the sea. This numerical landslide-tsunami hazard assessment technique for site-specific bathymetric and topographic conditions relies on the two complementing numerical codes DualSPHysics and SWASH; DualSPHysics is well-suited for the violent wave generation process, but is computationally expensive, and may result in unphysical wave decay beyond the coupling location. On the other hand, SWASH deals well with long-distance wave propagation and run-up by taking frequency dispersion and bottom friction into account at small computational cost. It was demonstrated that these two codes combined are capable of resolving the entire landslide-tsunami phenomenon from slide impact, wave generation, and wave propagation to run-up, as well as the inundation of coastal areas.

The landslide-tsunami scenarios investigated herein are hypothetical, and no evidence currently suggests that Es Vedrà may become unstable. Further, the predicted maximum wave amplitudes herein would be extreme values. This is because only subaerial landslides were investigated, which are known to generate larger tsunamis than partially submerged or submarine landslides by given slide volume. Further, extreme slide scenarios resulting in the largest waves were investigated, and it was in addition not possible to fully model the slide basal friction at this stage (the slides are rather modelled as hypermobile [59]). This issue has already been pointed out by Heller et al. [8], who reduced the slide impact velocity by approximately 50% to match the experimental wave amplitudes. In the present work, the calibration and validation tests of DualSPHysics v4.0 with the experimental data presented in Heller et al. [8] was once more performed, resulting in similar conclusions as described by Heller et al. [8], namely that the numerical wave amplitudes are overestimated with an unreduced slide impact velocity. More work is required to fully address this slide kinematics issue. In the meantime, it should be kept in mind that the simulated tsunamis are likely to be larger in the immediate slide impact zone than observed in reality. On the other hand, the wave decay in DualSPHysics was shown to be overpredicted in 3D by Heller et al. [8] (see their Figure 6e,f), such that the too-large slide velocity and wave decay partially compensate each other, and the simulated tsunami amplitudes at the coupling location are expected to be reliable enough for engineering applications. On the other hand, the wave propagation in SWASH is known to work reliably [48].

Finally, even though the wave propagation towards the mainland of Spain was not investigated herein, it is safe to state that a potential landslide tsunami originating from Es Vedrà would be very small at the mainland of Spain due to two main reasons. Firstly, Es Vedrà is less sloped towards the mainland of Spain, such that the slide volume and hence the tsunamis, would be smaller than in the two investigated scenarios. Secondly, the closest point on the mainland of Spain is 85 km away from Es Vedrà, and by applying the decay $r^{-1.19}$ that was found towards Cala d'Hort (which is likely to underpredict the decay of the more freely propagating tsunamis towards the mainland of Spain), the resulting wave would only be 0.16 m based on the largest wave amplitude of 133 m in scenario 1. Further, the affected coast of mainland Spain consists mainly of cliffs, which also reduces the damage potential of a hypothetical landslide tsunami from Es Vedrà.

5. Conclusions

This work addressed a numerical technique to conduct landslide-tsunami hazard assessments in reservoirs, lakes, fjords, and the sea. The viability of this technique was demonstrated with hypothetical landslide tsunamis originating from Es Vedrà, offshore Ibiza, under consideration of the site-specific bathymetry and topography. The two numerical codes DualSPHysics and SWASH were applied, thereby combining the strengths in modelling the violent wave generation process of the former with the accurate long-distance wave propagation at the small computational cost of the latter. The coupling of the two codes was carried out by importing the wave profiles and wave kinematics from DualSPHysics into SWASH at a distance where the wave profiles were reasonably stable. Two landslide-tsunami scenarios were investigated by focusing on two particularly exposed locations, namely (i) Cala d'Hort (3 km away from Es Vedrà) and (ii) Marina de Formentera (23 km away from Es Vedrà). The main findings of this study are summarised as follows:

— Two different slide–wave interaction phases were observed. (i) At the very beginning, the slide moved faster than the waves, such that the slide propagated below the primary wave crest and additionally elevated the water column and free water surface. (ii) The slide then slowed down such that the waves travelled faster and abruptly decayed due to the increased water depth.

— In scenario 1 (Cala d'Hort), the maximum wave amplitude was 133.0 m, reducing to a wave amplitude of 14.2 m at 5 m offshore and a maximum run-up height of over 21.5 m. In scenario 2 (Marina de Formentera), the maximum wave amplitude was 75.4 m, reducing to 2.7 m at 5 m offshore Marina de Formentera, such that the inundation depth was 1.2 m in the populated harbor area. This is significantly smaller than at Cala d'Hort, but may still result in significant devastation due to a larger density of buildings and infrastructure.

— The proposed numerical technique results likely in an overestimation of the landslide tsunamis because extreme slide scenarios were selected (extreme slide masses, slip orientation, and subaerial slides), and the slide velocity in DualSPHysics is likely to be overpredicted.

— The proposed numerical technique also provided new insights into 3D landslide-tsunami propagation by considering site-specific bathymetric and topographic conditions.

The change of the landslide-tsunami features by using the laminar viscosity with sub-particle scale turbulence rather than the artificial viscosity for the dissipative terms, the slide kinematics issue and the definition of an exact criterion for the coupling location remain open for future research.

Author Contributions: Conceptualisation, V.H.; Numerical simulations with DualSPHysics, H.T. with minor support from G.R. and V.H.; Numerical simulations with SWASH, G.R.; Analysis of data, H.T. (DualSPHysics) and G.R. (SWASH); Visualisation, G.R., H.T. and V.H.; Writing-Original Draft Preparation, H.T., apart from the sections involving SWASH which were written by G.R.; Writing-Review & Editing, G.R., H.T., S.C. and V.H.; Supervision, S.C. and V.H.; Project Administration, V.H.; Funding Acquisition, H.T., S.C. and V.H.

Funding: This research was funded by special funds for the postgraduate going abroad (outbound) exchange project of Wuhan University. The Titan Xp GPU used for this research was donated by the NVIDIA Corporation.

Acknowledgments: Hai Tan thanks Wuhan University for supporting him through the PhD Short-time Mobility Program to conduct research with Valentin Heller at the University of Nottingham, UK, from December 2017 to May 2018. Thanks go to Jiazhang Liu for initiating the DualSPHysics simulations of this work within his BEng individual investigative project. The Nottingham High Performance Computing (HPC) cluster Minerva has been accessed for the SWASH simulations.

Conflicts of Interest: The authors declare no conflict of interest.

Notation

a	Amplitude [m]
c	Speed of sound [m/s]
$\bar{c}$	Average speed of sound [m/s]
c_0	Speed of sound at the reference density [m/s]
c_f	Bottom friction coefficient [-]
C_r	Courant number [-]
d	Total water depth [m]
dp	Particle spacing [mm]
f	Field variable [-]
g	Gravitational acceleration [m/s^2]
$\mathbf{g}$	Gravitational acceleration vector [m/s^2]
h	Still water depth [m]
h_p	Smoothing length [m]
k	Wave number [-]
m	Mass [kg]
n	Manning's roughness coefficient [s/m$^{1/3}$]
p	Pressure [kg/ms^2]
p_h	Hydrostatic pressure [kg/ms^2]
p_t	Total pressure [kg/ms^2]
q	Non-hydrostatic pressure [kg/ms^2]
r	Radial distance from slide impact location [m]
t	Time [s]
$\bar{u}$	Depth-averaged velocity in x-direction [m/s]
v	Velocity vector [m/s]
$\bar{v}$	Depth-averaged velocity in y-direction [m/s]
v_{max}	Maximum flow velocity [m/s]
w_b	Vertical velocity at the bottom [m/s]
w_s	Vertical velocity at the free surface [m/s]
W	Weighting function or smoothing kernel [-]
x	Distance from the slide impact; coordinate along the slide axis [m]
$\mathbf{x}$	Position vector [m]
x_b	Distance from Cala d'Hort [m]
x_h	Distance from Marina de Formentera [m]
y	Coordinate perpendicular to the slide axis [m]
z	Coordinate vertical to the slide axis [m]
γ	Dimensionless parameter in the equation of state [-]
Γ	Dissipative term [-]
δ	Delta SPH coefficient [-]
Δt	Time step [s]
Δx	Grid resolution in the x-direction [m]
Δy	Grid resolution in the y-direction [m]
$\|\Delta\rho\|/\rho$	Relative density fluctuation [-]
ε	Dimensionless parameter in the XSPH variant [-]
η	Water surface elevation [m]
κ	Artificial viscosity coefficient [-]
μ	Intermediate variable in the artificial viscosity [-]
v_t	Horizontal eddy viscosity [m^2/s]
Π	Artificial viscosity [-]
ρ	Density [kg/m^3]
$\bar{\rho}$	Average density [kg/m^3]
ρ_0	Reference density [kg/m^3]
τ	Turbulent stress [kgm^3/s^2]
Ω	Computation domain [-]

J. Mar. Sci. Eng. **2018**, *6*, 111

Subscript

a, b	Fluid particles

Abbreviation

BC	Boundary Condition
CFL	Courant-Friedrichs-Lewy
CPU	Central Processing Unit
GPU	Graphics Processing Unit
HPC	High Performance Computing
MPS	Moving Particle Semi-implicit
NLSWE	Non-Linear Shallow Water Equation
PFEM	Particle Finite Element Method
RAM	Random Access Memory
SPH	Smoothed Particle Hydrodynamics
SWASH	Simulating WAve till SHore
SWL	Still Water Level
WCSPH	Weakly Compressible SPH

References

1. Panizzo, A.; Girolamo, P.D.; Risio, M.D.; Maistri, A.; Petaccia, A. Great landslide events in Italian artificial reservoirs. *Nat. Hazards Earth Syst. Sci.* **2005**, *5*, 733–740. [CrossRef]
2. Fuchs, H.; Pfister, M.; Boes, R.; Perzlmaier, S.; Reindl, R. Impulswellen infolge Lawineneinstoss in den Speicher Kühtai. *Wasserwirtschaft* **2011**, *101*, 54–60. [CrossRef]
3. Fuchs, H.; Boes, R. Berechnung felsrutschinduzierter Impulswellen im Vierwaldstättersee. *Wasser Energie Luft* **2010**, *102*, 215–221.
4. Gylfadóttir, S.S.; Kim, J.; Helgason, J.K.; Brynjólfsson, S.; Höskuldsson, Á.; Jóhannesson, T.; Harbitz, C.B.; Løvholt, F. The 2014 Lake Askja rockslide-induced tsunami: Optimization of numerical tsunami model using observed data. *J. Geophys. Res. Oceans* **2017**, *122*, 4110–4122. [CrossRef]
5. Harbitz, C.B.; Glimsdal, S.; Løvholt, F.; Kveldsvik, V.; Pedersen, G.K.; Jensen, A. Rockslide tsunamis in complex fjords: From an unstable rock slope at Åkerneset to tsunami risk in western Norway. *Coast. Eng.* **2014**, *88*, 101–122. [CrossRef]
6. Watt, S.F.L.; Talling, P.J.; Vardy, M.E.; Heller, V.; Hühnerbach, V.; Urlaub, M.; Sarkar, S.; Masson, D.G.; Henstock, T.J.; Minshull, T.A.; et al. Combinations of volcanic-flank and seafloor-sediment failure offshore Montserrat, and their implications for tsunami generation. *Earth Planet. Sci. Lett.* **2012**, *319*, 228–240. [CrossRef]
7. Vacondio, R.; Mignosa, P.; Pagani, S. 3D SPH numerical simulation of the wave generated by the Vajont rockslide. *Adv. Water Resour.* **2013**, *59*, 146–156. [CrossRef]
8. Heller, V.; Bruggemann, M.; Spinneken, J.; Rogers, B.D. Composite modelling of subaerial landslide–tsunamis in different water body geometries and novel insight into slide and wave kinematics. *Coast. Eng.* **2016**, *109*, 20–41. [CrossRef]
9. Zijlema, M.; Stelling, G. Efficient computation of surf zone waves using the nonlinear shallow water equations with non-hydrostatic pressure. *Coast. Eng.* **2008**, *55*, 780–790. [CrossRef]
10. Zijlema, M.; Stelling, G.; Smit, P. SWASH: An operational public domain code for simulating wave fields and rapidly varied flows in coastal waters. *Coast. Eng.* **2011**, *58*, 992–1012. [CrossRef]
11. Sælevik, G.; Jensen, A.; Pedersen, G. Experimental investigation of impact generated tsunami; related to a potential rock slide, Western Norway. *Coast. Eng.* **2009**, *56*, 897–906. [CrossRef]
12. Abadie, S.M.; Harris, J.C.; Grilli, S.T.; Fabre, R. Numerical modeling of tsunami waves generated by the flank collapse of the Cumbre Vieja Volcano (La Palma, Canary Islands): Tsunami source and near field effects. *J. Geophys. Res. Oceans* **2012**, *117*, C05030. [CrossRef]
13. Heller, V.; Spinneken, J. Improved landslide-tsunami prediction: Effects of block model parameters and slide model. *J. Geophys. Res. Oceans* **2013**, *118*, 1489–1507. [CrossRef]
14. Heller, V.; Hager, W.H. Impulse product parameter in landslide generated impulse waves. *J. Waterw. Port Coast. Ocean Eng.* **2010**, *136*, 145–155. [CrossRef]

15. Heller, V.; Hager, W.H. A universal parameter to predict landslide-tsunamis? *J. Mar. Sci. Eng.* **2014**, *2*, 400–412. [CrossRef]

16. Evers, F.M.; Hager, W.H. Spatial impulse waves: Wave height decay experiments at laboratory scale. *Landslides* **2016**, *13*, 1395–1403. [CrossRef]

17. Heller, V.; Hager, W.H.; Minor, H.-E. *Landslide Generated Impulse Waves in Reservoirs—Basics and Computation*; Swiss Federal Institute of Technology (ETH) Zurich: Zurich, Switzerland, 2009.

18. Battaglia, D.; Strozzi, T.; Bezzi, A. Landslide hazard: Risk zonation and impact wave analysis for the Bumbuma Dam-Sierra Leone. *Geol. Soc. Territory* **2015**, *2*, 1129–1134.

19. BGC Engineering Inc. *Appendix 4-E Mitchell Pit Landslide Generated Wave Modelling*; BGC Engineering Inc.: Vancouver, BC, Canada, December 2012.

20. Gabl, R.; Seibl, J.; Gems, B.; Aufleger, M. 3-D numerical approach to simulate the overtopping volume caused by an impulse wave comparable to avalanche impact in a reservoir. *Nat. Hazards Earth Syst. Sci.* **2015**, *15*, 2617–2630. [CrossRef]

21. Lüthi, M.P.; Vieli, A. Multi-method observation and analysis of a tsunami caused by glacier calving. *Cryosphere* **2016**, *10*, 995–1002. [CrossRef]

22. Oppikofer, T.; Hermanns, R.L.; Sandoy, G.; Böhme, M.; Jaboyedoff, M.; Horton, P.; Roberts, N.J.; Fuchs, H. Quantification of casualties from potential rock-slope failures in Norway. *Landslides Eng. Slopes. Exp. Theory Pract.* **2016**, 1537–1544.

23. Salazar, F.; Irazábal, J.; Larese, A.; Oñate, E. Numerical modelling of landslide-generated waves with the particle finite element method (PFEM) and a non-Newtonian flow model. *Int. J. Numer. Anal. Methods Geomech.* **2016**, *40*, 809–826. [CrossRef]

24. Tan, H.; Chen, S. A hybrid DEM-SPH model for deformable landslide and its generated surge waves. *Adv. Water Resour.* **2017**, *108*, 256–276. [CrossRef]

25. Kesseler, M.; Heller, V.; Turnbull, B. A laboratory-numerical approach for modelling scale effects in dry granular slides. *Landslides* **2018**, *15*, 1–15. [CrossRef]

26. Fu, L.; Jin, Y.C. Investigation of non-deformable and deformable landslides using meshfree method. *Ocean Eng.* **2015**, *109*, 192–206. [CrossRef]

27. Shi, C.; An, Y.; Wu, Q.; Liu, Q.; Cao, Z. Numerical simulation of landslide-generated waves using a soil–water coupling smoothed particle hydrodynamics model. *Adv. Water Resour.* **2016**, *92*, 130–141. [CrossRef]

28. Cremonesi, M.; Frangi, A.; Perego, U. A Lagrangian finite element approach for the simulation of water-waves induced by landslides. *Comput. Struct.* **2011**, *89*, 1086–1093. [CrossRef]

29. Heller, V.; Spinneken, J. On the effect of the water body geometry on landslide-tsunamis: Physical insight from laboratory tests and 2D to 3D wave parameter transformation. *Coast. Eng.* **2015**, *104*, 113–134. [CrossRef]

30. Crespo, A.J.C.; Domínguez, J.M.; Rogers, B.D.; Gómez-Gesteira, M.; Longshaw, S.; Canelas, R.; Vacondio, R.; Barreiro, A.; García-Feal, O. DualSPHysics: Open-source parallel CFD solver based on smoothed particle hydrodynamics (SPH). *Comput. Phys. Commun.* **2015**, *187*, 204–216. [CrossRef]

31. Violeau, D.; Rogers, B.D. Smoothed particle hydrodynamics (SPH) for free-surface flows: Past, present and future. *J. Hydraul. Res.* **2016**, *54*, 1–26. [CrossRef]

32. Madsen, P.A.; Fuhrman, D.R.; Schäffer, H.A. On the solitary wave paradigm for tsunamis. *J. Geophys. Res.* **2008**, *113*, C12012. [CrossRef]

33. Narayanaswamy, M.; Crespo, A.J.C.; Gómez-Gesteira, M.; Dalrymple, R.A. SPHysics-FUNWAVE hybrid model for coastal wave propagation. *J. Hydraul. Res.* **2010**, *48*, 85–93. [CrossRef]

34. Altomare, C.; Domínguez, J.M.; Crespo, A.J.C.; Suzuki, T.; Caceres, I.; Gómez-Gesteira, M. Hybridization of the wave propagation model SWASH and the meshfree particle method SPH for real coastal applications. *Coast. Eng. J.* **2015**, *57*, 1550024. [CrossRef]

35. Viroulet, S.; Cébron, D.; Kimmoun, O.; Kharif, C. Shallow water waves generated by subaerial solid landslides. *Geophys. J. Int.* **2013**, *193*, 747–762. [CrossRef]

36. European Marine Observation and Data Network, Bathymetry. Available online: http://portal.emodnet-bathymetry.eu/ (accessed on 15 May 2018).

37. CNIG, Topography. Available online: http://centrodedescargas.cnig.es/CentroDescargas/locale?request_locale=en (accessed on 15 May 2018).

38. Gomez-Gesteira, M.; Rogers, B.D.; Crespo, A.J.C.; Dalrymple, R.A.; Narayanaswamy, M.; Dominguez, J.M. SPHysics—Development of a free-surface fluid solver—Part 1: Theory and formulations. *Comput. Geosci.* **2012**, *48*, 289–299. [CrossRef]

39. Gomez-Gesteira, M.; Rogers, B.D.; Dalrymple, R.A.; Crespo, A.J.C. State-of-the-art of classical SPH for free-surface flows. *J. Hydraul. Res.* **2010**, *48*, 6–27. [CrossRef]

40. Tan, H.; Xu, Q.; Chen, S. Subaerial rigid landslide-tsunamis: Insights from a block DEM-SPH model. *Eng. Anal. Bound. Elem.* **2018**, *95*, 297–314. [CrossRef]

41. Liu, G.R.; Liu, M.B. *Smoothed Particle Hydrodynamics: A Meshfree Particle Method*, 1st ed.; World Scientific: Singapore, 2003.

42. Molteni, D.; Colagrossi, A. A simple procedure to improve the pressure evaluation in hydrodynamic context using the SPH. *Comput. Phys. Commun.* **2009**, *180*, 861–872. [CrossRef]

43. Monaghan, J.J. Smoothed particle hydrodynamics. *Rep. Prog. Phys.* **2005**, *68*, 1703–1759. [CrossRef]

44. Monaghan, J.J. Smoothed particle hydrodynamics. *Annu. Rev. Astron. Astrophys.* **1992**, *30*, 543–574. [CrossRef]

45. Monaghan, J.J. Simulating free surface flows with SPH. *J. Comput. Phys.* **1994**, *110*, 399–406. [CrossRef]

46. Morris, J.P.; Fox, P.J.; Zhu, Y. Modeling low Reynolds number incompressible flows using SPH. *J. Comput. Phys.* **1997**, *136*, 214–226. [CrossRef]

47. Crespo, A.J.C.; Gómez-Gesteira, M.; Dalrymple, R.A. Boundary conditions generated by dynamic particles in SPH methods. *CMC Comput. Mater. Contin.* **2007**, *5*, 173–184.

48. Ruffini, G.; Heller, V.; Briganti, R. Numerical modelling of landslide-tsunami propagation in a wide range of idealised water body geometries. 2018. (in preparation)

49. Stelling, G.; Zijlema, M. An accurate and efficient finite-difference algorithm for non-hydrostatic free-surface flow with application to wave propagation. *Int. J. Numer. Methods Fluids* **2003**, *43*, 1–23. [CrossRef]

50. Zijlema, M.; Stelling, G. Further experiences with computing non-hydrostatic free-surface flows involving water waves. *Int. J. Numer. Methods Fluids* **2005**, *48*, 169–197. [CrossRef]

51. Blayo, E.; Debreu, L. Revisiting open boundary conditions from the point of view of characteristic variables. *Ocean Model.* **2005**, *9*, 231–252. [CrossRef]

52. Sommerfeld, A. Die Greensche Funktion der Schwingungsgleichung. *Jahresber. Dtsch. Mathematiker-Ver.* **1912**, *21*, 309–353.

53. *SWASH—User Manual*; Version 4.01; Delft University of Technology: Delft, The Netherland, 2016.

54. Nehlich, O.; Fuller, B.T.; Márquez-Grant, N.; Richards, M.P. Investigation of diachronic dietary patterns on the islands of Ibiza and Formentera, Spain: Evidence from sulfur stable isotope ratio analysis. *Am. J. Phys. Anthropol.* **2012**, *149*, 115–124. [CrossRef] [PubMed]

55. Manger, G.E. *Porosity and Bulk Density of Sedimentary Rocks*; U.S. G.P.O.: Washington, DC, USA, 1963.

56. Hughes, S.A. *Physical Models and Laboratory Techniques in Coastal Engineering*, 1st ed.; World Scientific: Singapore, 1993.

57. Heller, V. Scale effects in physical hydraulic engineering models. *J. Hydraul. Res.* **2011**, *49*, 293–306. [CrossRef]

58. Wang, J.; Ward, S.N.; Xiao, L. Numerical simulation of the December 4, 2007 landslide-generated tsunami in Chehalis Lake, Canada. *Geophys. J. Int.* **2015**, *201*, 372–376. [CrossRef]

59. Pudasaini, S.P.; Miller, S.A. The hypermobility of huge landslides and avalanches. *Eng. Geol.* **2013**, *157*, 124–132. [CrossRef]

60. Huang, B.; Yin, Y.; Chen, X.; Liu, G.; Wang, S.; Jiang, Z. Experimental modeling of tsunamis generated by subaerial landslides: Two case studies of the Three Gorges Reservoir, China. *Environ. Earth Sci.* **2014**, *71*, 3813–3825.

© 2018 by the authors. Licensee MDPI, Basel, Switzerland. This article is an open access article distributed under the terms and conditions of the Creative Commons Attribution (CC BY) license (http://creativecommons.org/licenses/by/4.0/).

Journal of
Marine Science and Engineering

MDPI

Article

Case Study of Dam Overtopping from Waves Generated by Landslides Impinging Perpendicular to a Reservoir's Longitudinal Axis

Netsanet Nigatu Tessema [1],*, Fjóla G. Sigtryggsdóttir [2], Leif Lia [2] and Asie Kemal Jabir [1]

[1] School of Civil and Environmental Engineering, Addis Ababa Institute of Technology, P. O. Box 385, Addis Ababa 1000, Ethiopia

[2] Department of Civil and Environmental Engineering, Norwegian University of Science and Technology, 7491 Trondheim, Norway

* Correspondence: Netsanet.nigatu@aait.edu.et; Tel.: +251-913-77-15-74

Received: 29 May 2019; Accepted: 25 June 2019; Published: 15 July 2019

Abstract: Landslide-generated impulse waves in dammed reservoirs run up the reservoir banks as well as the upstream dam slope. If large enough, the waves may overtop and even breach the dam and cause flooding of the downstream area with hazardous consequences. Hence, for reservoirs in landslide-prone areas, it is important to provide a means to estimate the potential size of an event triggered by landslides along the reservoir banks. This research deals with landslide-generated waves and the overtopping process over the dam crest in a three-dimensional (3D) physical model test, presenting a case study. The model set-up describes the landslide impacting the reservoir in a perpendicular manner, which is often the case in natural settings. Based on the experimental results, dimensionless empirical relations are derived between the overtopping volume and the governing parameters, namely the slide volume, slide release height, slide impact velocity, still-water depth, and upstream dam face slope. Predictive relations for the overtopping volume are presented as applicable for cases relating to the specific model set-up. Measured overtopping volumes are further compared to a two-dimensional (2D) case reported in the literature. An important feature regarding the overtopping process for the 3D case is the variation in time and space, resulting in an uneven distribution of the volume of water overtopping the dam crest. This observation is made possible by the 3D model set-up, and is of value for dam safety considerations as well as for foundation-related issues, including erosion and scouring.

Keywords: landslide-generated wave; dam overtopping; physical model; overtopping volume; impulse wave

---

1. Introduction

Dam sites suitable for impounding of a reservoir are often found in mountainous regions or highlands, often in narrow valleys or canyons (see Figure 1). Mountain slopes are generally susceptible for landslides, including rockslides. There are infamous cases of a landslide impinging reservoirs, in turn generating impulse waves overtopping the dam with catastrophic consequences downstream. This includes the Vajont dam tragedy in 1963 where nearly 2000 fatalities occurred [1]. The general process describing such events has been grouped into three phases [2]: (1) slide impact with wave generation; (2) wave propagation with wave transformation; and (3) run-up of the impulse wave and overtopping of a dam (see Figure 2). However, a dam is not overtopped if the wave run-up height is lower than the freeboard f, the elevation difference between the dam crest, and the reservoir still-water level, when the landslide impinges. Still, mountainous slopes surrounding a dam reservoir poses a threat to dam safety. Consequently, a mean of estimating the associated hazard from landslide waves overtopping a dam is of importance. In such an evaluation, the reservoir settings, such as the geology and topography, are important. The shape of a reservoir in a narrow mountain valley is usually longer than its width (see Figure 1), and thus with a potential landslide threat from the mountain slopes along the length of the reservoir. In other words, a potential landslide may fall from these mountain slopes, i.e., approximately perpendicular to the reservoir's longitudinal axis (see e.g., Figure 2b).

Figure 1. Mountain reservoir in Aurland Municipally Norway (Photo from E-CO Energi / Aerosport, with permission, 2019).

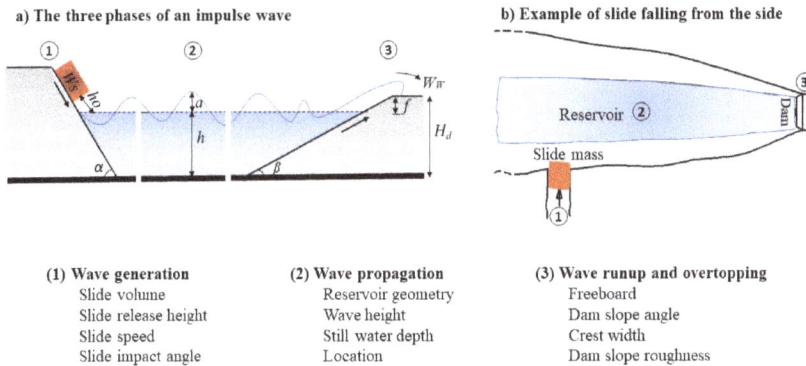

a) The three phases of an impulse wave

b) Example of slide falling from the side

(1) Wave generation	(2) Wave propagation	(3) Wave runup and overtopping
Slide volume	Reservoir geometry	Freeboard
Slide release height	Wave height	Dam slope angle
Slide speed	Still water depth	Crest width
Slide impact angle	Location	Dam slope roughness

Figure 2. The three phases of landslide-generated waves with the relevant parameters for this study: (1) slide impact with wave generation; (2) wave propagation; and (3) wave run-up and overtopping of a dam; (**a**) a section showing the three phases based on Heller et al. [2]; (**b**) an example plan view of a reservoir.

Numerous experimental and numerical studies on landslide-generated waves (Phase 1 in Figure 2) are available. Many of those were conducted in flumes to investigate two-dimensional (2D) properties [3–7], while other studies [8–17] investigated the three-dimensional (3D) effects of landslide-generated tsunamis by considering wide reservoirs (3D water bodies) [11,16,17], as well as several geometries such as planar beaches and islands [8,12–14]. Those studies include landslides modeled with granular and cobble material, and investigation into the effect of different landslide parameters such as landslide geometry and energy, on the wave generated. One recent relevant study is that of Evers [16], who investigated, for example, the effect of slide impact velocity, slide mass, slide impact angle, slide width, and still-water depth in 2D and 3D models. Evers [16] found that for the same slide mass (volume) the wave generation is influenced by the landslide geometry, with a wider landslide generating larger waves. Furthermore, he found that the initial wave amplitude and shape are influenced by the slide impact angle. These findings are mentioned here since in the current study the geometry of the slide is not directly considered in the formulations provided, and the slide impact angle is constant.

Only a few studies consider dam overtopping (Phase 3 in Figure 2) and include formulas for the overtopping volume based on wave and dam parameters. These studies are mainly those of Huber et al. [17] in the case of an erodible granular dam model, and Kobel et al. [18] as well as Müller [19] in the case of a solid, non-erodible dam model. The resulting formula for the overtopping volume is based on (2D) experimental tests. Kobel et al. [18] used rectangular prismatic water wave channels with solitary waves propagating directly towards a dam (set-up similar to shown in Figure 3b), thereby simulating waves generated by a landslide impinging from one end of the reservoir directly towards the dam. Kobel et al. [18] point out that the solitary wave type represents an extreme case.

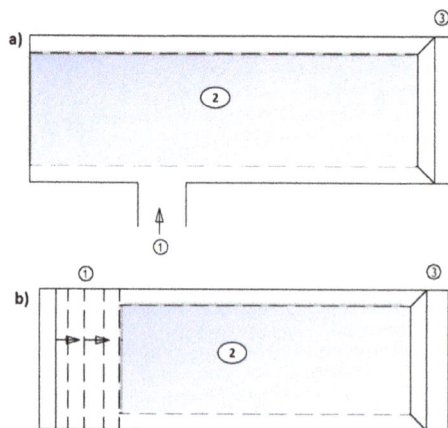

Figure 3. General layout of slide impact into a reservoir: 1 is the slide impact zone, 2 is the wave propagation zone and 3 the wave run-up and dam overtopping zone for (**a**) slide impacting from the side of a gorge; (**b**) solitary wave generated moving directly towards the dam's upstream face.

Depending on the geometry of a particular reservoir, the wave parameters can be computed with a 3D or 2D approach (e.g., [11,16]) and fed into a 2D run-up equation, such as provided by Kobel et al. [18], to obtain an estimate of the volume of water overtopping a dam. Heller et al. [2] discuss the effect of the reservoir shape with two extreme cases. The first case considers a long reservoir and a slide that impacts longitudinally into this reservoir. The second case considers that a slide mass can impact at any possible location into the reservoir and the slide width is less than the reservoir width. In the second case the reservoir geometry is such that the impulse wave can propagate radially and freely from the slide impact zone. In 3D settings the wave parameters depend on the wave propagation angle. The distribution of the overtopping wave along the dam crest, e.g., at the inner and outer flanks (see e.g., [2]), can be considered by dividing the dam crest in appropriate number of sections for the calculation of wave run-up and subsequent overtopping. However, the quality of the prediction would depend on the geometry of the reservoir and may be limited in the case of a very narrow reservoir as in the present study.

The above-mentioned studies do not directly investigate the 3D effect relating to narrow valleys and a landslide falling into a narrow dammed reservoir perpendicular to the reservoir's longitudinal axis, i.e., setups related to this shown in Figure 3a. Thus, further studies are required to investigate directly dammed reservoirs in narrow valleys with mountain slopes prone to landslide impinging perpendicular to the longitudinal axis of the reservoir (Figure 3a). The current study aims at shedding light on the mentioned 3D effect; however, application of the results must consider the limitations of the model set-up of this case study, as later described.

The current study continues an experimental study program initiated in 2014 with focus on landslide-generated waves resulting from lateral slides into a narrow reservoir, as well as the effect of different dam related parameters on wave overtopping. The study uses a 3D physical model that was extracted and modified from a model used by Lindstrøm et al. [20], to study rockslide generated waves into fjords. The same slide blocks are used as in Lindstrøm et al. [20], i.e., essentially modeling rockslides. Other researchers have studied landslide-generated impulse waves using granular deformable slides, e.g., Fritz et al. [21,22] used a 2D physical model and Mohammed and Fritz [23] used a 3D tsunami wave basin. Furthermore, there are studies available, e.g., Ataie-Ashtiani and Nik-Khah [7], Zweifel [24] and Heller and Spinneken [25], comparing the waves generated by granular and block slides. The findings of Zweifel [24] and Ataie-Ashtiani and Nik-Khah [7] was that rigid blocks (as in the present study), result in more extreme waves compared to granular slides. However, Heller and Spinneken [25] revealed that block slides can generate not only larger waves in a wave channel, but also identical or smaller waves than granular slides, depending on certain parameters or features. The identified influential parameters in this regard were the ratio of the slide width to the channel width, the slide front angle and the slope transition from the inclined landslide ramp used in the tests to the horizontal channel bottom. They explain the discrepancy from the previous studies that these parameters were not varied, and their effect thus not recognized.

The main objective of this article is to use the experimental data from the 3D laboratory scale test set-up to propose a formula for estimating the total volume of water overtopping a dam from impulse waves generated by a landslide impinging perpendicular to the reservoir longitudinal axis as shown in Figure 3a. An important part of this objective is to investigate the distribution of the overtopping volume along the dam crest, a feature that can, as described above, only be obtained indirectly through 2D modeling of previous studies essentially using experimental setups relating to that shown in Figure 3b. The experimental set-up of the present study has a fixed reservoir geometry as well as a fixed landslide ramp slope and location, which must be considered in the application of the results. The formula derived is dependent on both the landslide and dam parameters. The dam parameters considered, for the upstream slope and freeboard, relate e.g., to those relevant for embankment dams. However, the dam model used is non-erodible as in the case of Kobel et al. [18]. Considering this, an important further objective of this study is to compare, the overtopping volume measured in this study to predictions based on the formula of Kobel et al. [18].

In the following, the experimental set-up, instrumentation, and test program is outlined, followed by a description of the wave propagation and overtopping process. The experimental data is analyzed to obtain a formula for predicting the overtopping volume based on the landslide and dam parameters. The formula is obtained with data regression analysis resulting from test runs with systematic variation of slide, basin, and dam parameters such as the slide volume, slide release height, slide speed, wave amplitude and upstream dam face slope for cases with freeboard $f > 0$. Following this, the distribution of the overtopping along the dam crest is extracted. Finally, the measured overtopping volumes are compared to results using the prediction formula proposed by Kobel et al. [18]. The analysis chapter is supported with discussions on the limitations of the study and the results. Finally, the main conclusions are stated.

2. Methodology

2.1. Experimental Set-up

The main structural components of the physical model were: (i) a slide ramp on which slide blocks are released; (ii) the basin representing the reservoir; and (iii) an embankment dam (non-erodible) (Figure 4).

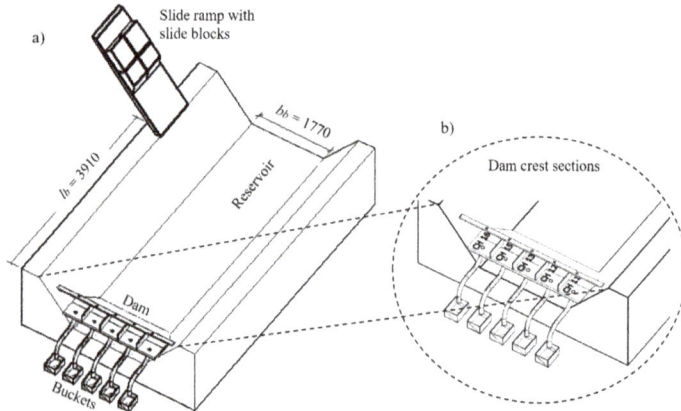

Figure 4. (**a**) Experimental set-up with the main components; slide, reservoir, dam and (**b**) ultrasonic sensors used for measuring the run-up height and buckets for collecting the overtopping water. The wave gauges, which were installed in the reservoir, are not shown on this figure.

The slide was modeled with blocks (see Figure 5) placed on a 2 m long slide ramp inclined at a constant angle of $\alpha = 50°$, where it was possible to place rectangular blocks of different sizes and arrangements. This model represents a reservoir in a narrow valley formed by steep rock mountain slopes. Furthermore, the slide ramp was modeled as a natural extension of the sides of the model reservoir. The inclination of the slide ramp mainly affects the slide speed, along with the friction between the ramp and the blocks. The friction resistance was represented with a friction angle of about 25°. In this study, the slide speed was varied by using different slide release heights. The shapes and sizes of the blocks used in the experiments are shown in Figure 5. Different slide block arrangements were made for each test set-up. For each arrangement the blocks, impinging the water first, were tapered at an angle of 45° at the front to simulate a slide that has a smaller front and a larger body [26]. The slide blocks were attached to each other with chains, the sliding body was then attached to a steel panel with a hook. When the hook was removed, the blocks slid into the reservoir generating impulsive waves.

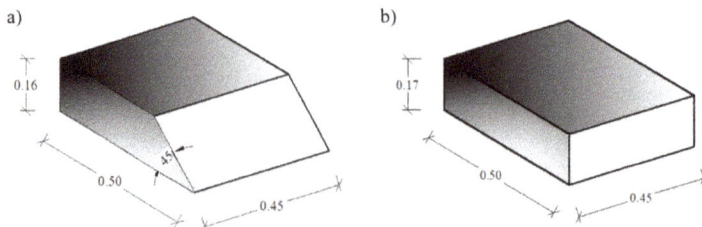

Figure 5. Rigid blocks used to model the landslide (measurement in m): (**a**) trapezoidal solid; (**b**) rectangular solid.

The reservoir was modeled with a fixed trapezoidal basin with sidewalls of water-resistant plywood covered with concrete paste to increase the roughness. It was 4.5 m long, 1.7 m wide at the bottom and 2.2 m at dam crest level, with a total reservoir capacity of 2.5 m³. The model dam was placed at one end of the reservoir with a constant height of $H_d = 0.32$ m. The model was conceptual and a scale of 1:190 has been selected relative to prototype, considering that this results in a moderate height of the prototype dam (60.8 m) and a range of landslide volumes from the blocks, representing medium sized slides (0.2 to 1 million m³) to large slides (1 to 5 million m³). At this scale (1:190) the model represents an 860 m long reservoir, 320 m wide at the bottom with a reservoir capacity of 17 Mm³.

Dams with upstream slopes of 1:1.5, 1:2 and 1:2.25 were used for the analysis representing embankment dams. The dam crest was divided into five different sections (labeled Channels (CH) 11 to 16 in Figure 4b) to measure the distribution of the overtopping volumes along the crest. The corresponding overtopping volume for each dam section was collected in five buckets with pipes of 100 mm diameter (see Figure 4b).

2.2. Instrumentation

Nine wave gauges of type 'DHI wave-meter 102E' were placed at the surface of the trapezoidal basin to measure the wave height at different locations. The principle of this wave meter is to measure the conductivity between two parallel electrodes partly immersed in water. In addition, there were five ultrasonic sensors (*mic+35/IU/TC*) placed at the top of each dam section to measure the height of the water that overtopped the dam crest (see Figure 6). The water that overtopped the dam was collected in five calibrated buckets and recorded manually with an ultrasonic sensor. The initial level of each bucket was measured before the test and the water level rise after the test. The measurement was multiplied by a calibration factor to obtain the overtopping volume.

To measure the speed of the slide block, a rotational sensor (CH 10) was placed at the side of the slide ramp. It was connected to the rigid sliding blocks through a hook with a rope which unrolled together with the distance covered by the slide. A voltmeter recorded the voltage as the block slid down the ramp and the rope unrolled. A calibration factor was required to change the measured voltage into the distance covered by the slide. Then a time variable distance was used to extract the slide impact velocity considering the impact velocity is the maximum.

The data from the wave gauges, ultrasonic sensors and rotational sensors were collected in 'Agilent Measuring Manager program' with a sampling rate of 200 Hz.

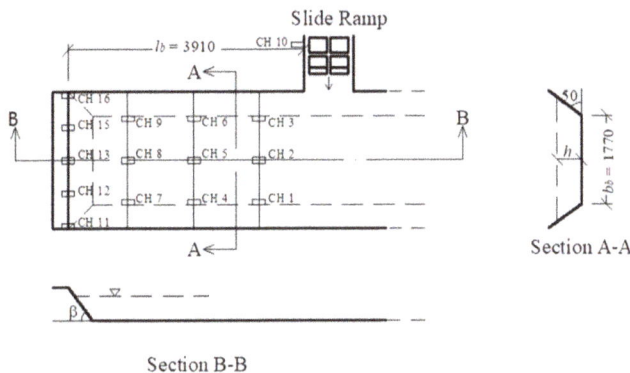

Figure 6. Planar view of sensors placement in the model set-up (measurements in mm).

2.3. Test Program

The overtopping of the dam by landslide-generated waves was investigated with parameters such as the landslide volume W_S, landslide release height h_O, landslide speed v_S, upstream dam slope

β and freeboard f. These parameters represent the input variables of the study provided as shown in Table 1. Four different types of block arrangements (2H, 2V, 4 and 6) with different length, width, and volume were used in the model as shown in Table 2 (Figure 7).

Table 1. Test program.

Scale	β (V:H)	f (m)	W_S (m^3)	h_O (m)
Model (1:190)	1: 2.25	0.024, 0.032	0.072, 0.074, 0.149, 0.225	0.5, 1, 1.5, 2
	1: 2	0.024, 0.032	0.072, 0.074, 0.149, 0.225	0.5, 1, 1.5, 2
	1: 1.5	0.024, 0.032	0.072, 0.074, 0.149, 0.225	0.5, 1, 1.5, 2
Prototype	-	4.5, 6	0.49, 0.51, 1.02, 1.54 (Mm^3)	95, 190, 285, 380

Table 2. Slide block characteristics in the model set-up.

Slide Characteristics	Block Arrangement			
	2H Blocks	2V Blocks	4 Blocks	6 Blocks
Slide length l_S (m)	0.50	1.08	1.08	1.66
Slide width b (m)	0.90	0.45	0.90	0.90
Shape ratio l_S/b (-)	0.56	2.40	1.20	1.84
Slide volume W_S (m^3)	0.072	0.074	0.149	0.225

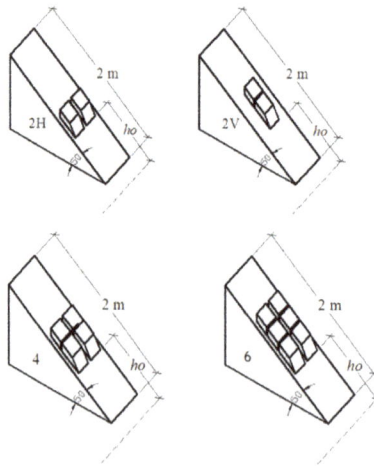

Figure 7. Arrangement of block configurations used in the tests with slide release height h_o parallel to the ramp slope and measured from the still-water level.

A constant dam height $H_d = 0.32$ m, crest width $b_c = 0.053$ m and length $l_c = 2.2$ m were used for the whole test series. However, three different upstream dam slopes were considered, i.e., $\beta = 24°$ (1:2.25), 27° (1:2) and 34° (1:1.5) (see Figure 8). Additionally, two different freeboards were used: $f = 0.024$ m (still-water depth: $h = 0.296$ m) and 0.032 m ($h = 0.288$ m). Considering a scale of 1:190, the freeboard values correspond to 4.5 m and 6 m, respectively, associated with dams in Norway of high and very high consequences for the downstream area should the dam breach.

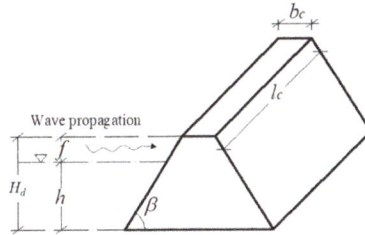

Figure 8. Definition sketch of a dam with the governing parameters.

66 experiments were conducted with varying slide, reservoir, and dam parameters. To check the test repeatability, each test with identical parameters was repeated three times and analyzed, hence a total of 198 tests have been conducted. All test data are given as the average of these three individual tests.

3. Wave Types, Propagation, and Overtopping Process

3.1. Wave Types and Propagation

Heller et al. [2] observed four transient impulse wave types for their 2D experiments in a wave channel: Stokes-like waves, cnoidal-like waves, solitary-like waves, bore-like waves, and all landslide-generated impulse waves may be allocated to one of these types. Each wave type (see Figure 9) has its own characteristics. The wave type influences for example the run-up height and wave force on the dam (see e.g., Heller and Hager [27]). Figure 10 provides an example of the wave profiles from all nine wave gauges on the reservoir, recorded during one of the 3D tests of this study (Test no. 185_2.25s_4.5_2H_200). The 3D waves generated in this study are more complex than in the 2D case; however, compared to the profiles of Figure 9, the recorded waves from all the tests conducted are Stokes-like waves, rather than any of the other wave types. Furthermore, investigation into selected waves revealed a ratio of wavelength to reservoir depth in the range of intermediate water waves.

A classification method to characterize landslide-generated waves is of interest and available in the literature. Heller et al. [2] identified the wave type product, $T = S^{1/3}M \cos[(6/7)\alpha]$, as a relevant number in this regard, where M is the relative slide mass ($M = m_s/(\rho_w bh^2)$, with m_s as the slide mass and ρ_w as the density of water) and S is the relative slide thickness ($S = s/h$, with s as the slide thickness). Previously, Wiegel et al. [3] and Noda [28] defined the slide Froude number F as the ratio of the falling box velocity v_s, and the shallow water waves celerity ($F = v_s/\sqrt{gh}$) and used F to identify the wave types produced by a falling block. Considering these numbers, Heller and Hager [27] defined wave type zones from the relationship between F and T for the 2D case. The boundaries of these wave type zones are plotted in Figure 11, where the axes show T versus F. Heller and Hager [27] observed for the 2D case mainly Stokes-like waves in the zone $T < 4/5F^{-7/5}$, mainly bore-like waves in the zone $T > 11F^{-5/2}$, and mainly cnoidal- and solitary-like waves in the intermediate zone $11F^{-5/2} \leq T \leq 4/5F^{-7/5}$. It is important to note that Heller and Hager [27] conducted granular slide tests, which may lead to different wave types under identical dimensionless parameters when compared to rigid slides such as in the present study. Heller and Spinneken [11] further investigated the wave types for the 3D case. They concluded from their 3D test results that the lower boundary ($T < 4/5F^{-7/5}$) remains characteristic also for 3D waves, whereas in the intermediate range both Stokes- and cnoidal-like waves were observed. Similar findings can be concluded from the 66 data points from the current 3D study plotted in Figure 11 and identified by the number of blocks used in the tests. The waves in all the tests conducted in the present study bear a resemblance to Stokes-like waves. This also applies to tests, involving 6-block as well as some of the 4-block tests that are plotted in Figure 11 in the intermediate zone defined for the 2D case for cnoidal- or solidary-like waves.

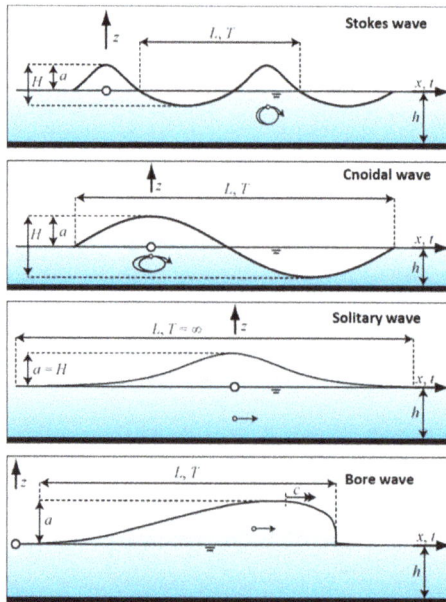

Figure 9. Idealized impulse wave types with the most important wave parameters (from Heller et al. [2], with permission from ETH, 2019).

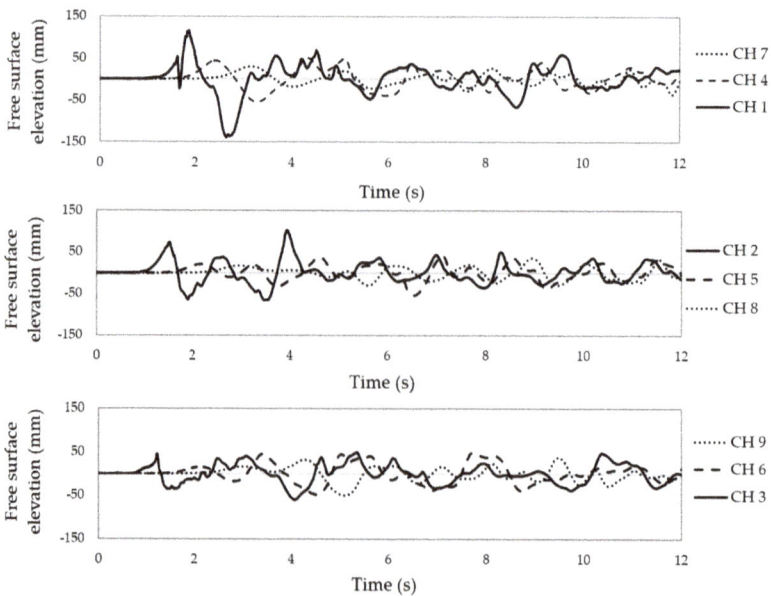

Figure 10. Wave profiles measured with the wave gauge sensors (see Figure 6 for the locations of the sensors) (Test no. 185_2.25s_4.5_2H_200).

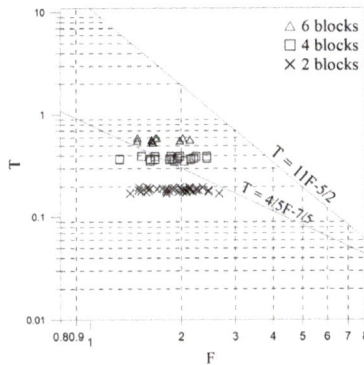

Figure 11. Plot of the wave type product T versus the Froude number F for the 3D tests conducted in the present study. All waves observed were Stokes-like. The boundaries provided are from Heller and Hager [27] observed for the 2D case using granular slides.

3.2. Wave Overtopping Process

Impulse waves induced by landslides impacting into reservoirs propagate and create wave run-up on the dam and shorelines. If the induced wave run-up exceeds the dam freeboard f the water overtops the dam and floods the downstream area. Kobel et al. [18] give a detailed description of the overtopping process for the 2D case and investigate the effect of the upstream slope, the freeboard and the dam width. Here, considering a 3D model, the focus is on variation in time and space of the overtopping waves, and thus the overtopping volume. During the overtopping process, the highest waves force a large volume of water over the crest in a short period of time, whereas the smaller waves may not produce any overtopping. The variation of the overtopping process in both time and space can be understood from Figure 12 presenting the time series obtained from all the sensors at the channels along the dam crest.

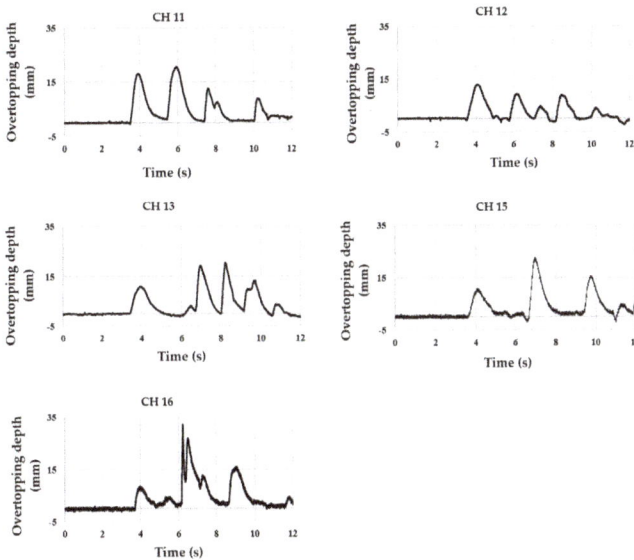

Figure 12. Overtopping depth (mm) versus time (s) over the five dam crest sections (see Figure 6 for the locations of the sensors) (Test no. 185_2.25s_4.5_2H_200).

4. Experimental Results

The experimental results are presented in this section. First, equations for the overtopping volume are derived for the experimental results. Furthermore, the uneven distribution of the overtopping volume is addressed.

4.1. Overtopping Volume

Once impulse waves are generated by landslides, they propagate and if large enough, overtop the dam thereby creating a certain volume of overtopping water. In this section, the relationship between the dimensionless parameters for the overtopping process will be assessed using an approach relating to e.g., that of Kobel et al. [18] and a predictive equation will be presented for the overtopping volume.

A dimensional analysis was conducted between the overtopping volume and the independent variables considering different landslide, reservoir, and dam geometries. Hence, the overtopping volume over the dam crest due to landslide-generated waves can be expressed as a function of:

$$W_w = f(W_s, h_o, g, v_s, h, \rho_w, a, l_b, b_b, \beta) \tag{1}$$

where W_w (m^3) = overtopping volume, W_s (m^3) = landslide volume, h_o (m) = landslide release height, g (m/s^2) = gravitational acceleration, v_s (m/s) = slide impact velocity, h (m) = still-water depth, a (m) = wave amplitude, l_b (m) = reservoir length, b_b (m) = reservoir width and β (°) = dam front face angle,.

In the model experiments a constant reservoir length, l_b and reservoir width, b_b were used; hence their effect is considered constant.

Applying Buckingham's π theorem, by selecting the repeating variables as; h, ρ_w and v_s, a relationship between dimensionless parameters is obtained:

$$\frac{W_w}{h^3} = f\left(\frac{W_s}{h^3}, \frac{h_o}{h}, \frac{v_s}{\sqrt{gh}}, \frac{a}{h}, \beta/90^0\right) \tag{2}$$

where the wave amplitude a is measured from the zero level to the maximum crest point. Equation (2) can be rewritten:

$$\frac{W_w}{h^3} = f\left(\frac{W_s}{h^3}, \frac{h_o}{h}, F, \varepsilon, \beta/90^0\right) \tag{3}$$

where $\varepsilon = a/h$ is the relative wave amplitude.

Conducting a power fit regression, a predictive equation is found for the relative overtopping volume:

$$\frac{W_w}{h^3} = 0.21\left[\left(\frac{W_s}{h^3}\right)\left(\frac{h_o}{h}\right)^{0.43}(F^2)^{-0.08}\varepsilon^{0.04}(\beta/90^0)^{-0.01}\right] = 0.21E_1 \tag{4}$$

The following limitations apply: $2.67 < W_s/h^3 < 9.52$, $1.67 < h_o/h < 6.97$, $1.25 < F < 2.66$, $0.18 < \varepsilon < 0.73$ and $0.27 < (\beta/90^0) < 0.37$.

E_1 is the overtopping volume parameter considering the relative wave amplitude described as:

$$E_1 = \left[\left(\frac{W_s}{h^3}\right)\left(\frac{h_o}{h}\right)^{0.43}(F^2)^{-0.08}\varepsilon^{0.04}(\beta/90^0)^{-0.01}\right]; \quad 2.9 < E_1 < 14.8 \tag{5}$$

Equation (4) includes a parameter relating to the wave crest; however, it also includes the relative slide volume which is influential for the wave amplitude as well as the landslide speed through F. However, the impact of ε, as well as $\beta/90^0$ and F, is quite small, indicating that these parameters can be neglected in the analysis of the overtopping volume W_w. The model set-up had a fixed distance from the dam to the slide impact zone. Thus, for this case a more direct relationship with the slide volume is possible by neglecting the wave parameter ε. Rearranging the parameters in Equation (4), for a direct relationship between W_w and the other parameters on the right-hand side of the equation, results in a relationship for W_s, to the power of 1. Thus, the relationship between W_w and W_s is linear. For cases that can be related to the model set-up, and where no information about the wave properties are available, removal of the wave related parameter is advantageous for a clearer extraction of the relationship between overtopping volume and slide parameters. This further simplifies the process of roughly estimating the overtopping volume related to a known potential slide into a reservoir. Hence, in this study, an additional analysis was made for a predictive equation of overtopping volume excluding wave properties, such as ε. Accordingly, the overtopping volume prediction formula becomes

$$\frac{W_w}{h^3} = 0.17\left[\left(\frac{W_s}{h^3}\right)\left(\frac{h_0}{h}\right)^{0.42}\left(F^2\right)^{-0.03}\left(\beta/90^0\right)^{-0.1}\right] = 0.17E_2 \tag{6}$$

The following limitations apply: $2.67 < W_s/h^3 < 9.52$, $1.67 < h_0/h < 6.97$, $1.25 < F < 2.66$ and $0.27 < \left(\beta/90^0\right) < 0.37$

E_2 is the overtopping volume parameter without considering ε described as:

$$E_2 = \left[\left(\frac{W_s}{h^3}\right)\left(\frac{h_0}{h}\right)^{0.42}\left(F^2\right)^{-0.03}\left(\beta/90^0\right)^{-0.1}\right]; \ 3.7 < E_2 < 17.8 \tag{7}$$

Based on the analysis the relative overtopping volume W_W/h^3 increases linearly and thus most significantly with the relative slide volume W_S/h^3, but moderately with the relative landslide release height h_0/h. Furthermore, the impacts of F and $\beta/90^0$ are again quite small. The exponents for F and $\beta/90^0$ in Equation (6) are so small that they may be ignored in the analysis of the overtopping volume. When excluding these parameters, the exponent of the remaining parameters, W_s/h^3 and h_0/h, remains the same for W_s/h^3, but only slightly reduces for h_0/h from 0.42 to 0.40. Furthermore, the constant in front of E_2 increases from 0.17 to 0.19. However, in the present study Equation (6) is considered for further analysis.

Figure 13 shows the correlation between measured relative overtopping volume W_W/h^3 and the overtopping volume parameters, E_1 and E_2, each with $R^2 = 0.80$. To investigate and compare Equation (4) including, and Equation (6) excluding the relative wave amplitude, an analysis was performed for the value of overtopping volume comparing the power regression fits (expressed by $R^2 = 0.80$) between the measured values from the tests and the results from Equations (4) and (6), respectively. Figure 14 reveals large prediction errors for smaller measured overtopping volumes $W_W < 0.01$ m^3 for both cases.

4.2. Overtopping Volume Distribution

Overtopping of a dam due to landslide-generated impulse waves is not uniform over the crest, for the case studied here with the landslide impinging along the reservoir's longitudinal axis (Figures 3a and 4). The water overtops the left edge first, and then the right edge and lastly the whole dam crest length. Figure 15 presents the distribution of the overtopping volume along the dam crest showing that a large amount of water is collected at the right and left edges (CH 16 and CH 11, respectively) of the crest, on average about 28% each. On average a slightly higher volume is collected at the left side, i.e., opposite the landslide impact zone. Contrariwise, a smaller volume of water overtops the middle section of the dam crest (CH 12, CH 13 and CH15) (Figure 15). This uneven distribution of the overtopping wave and volume along the crest of the dam is due to the 3D modeling allowing reflected waves from the edges of the reservoir.

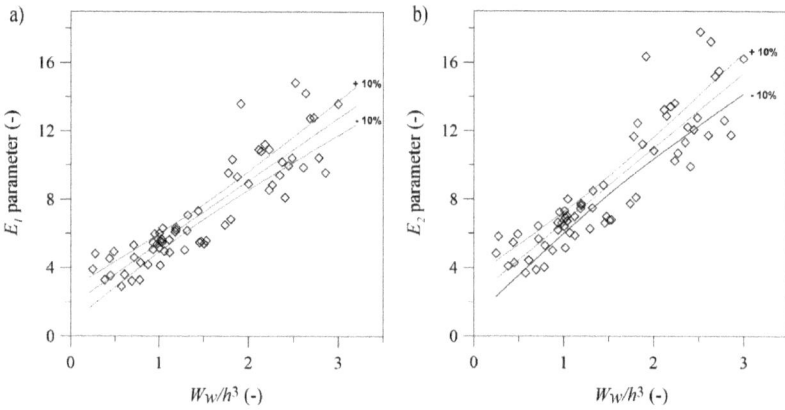

Figure 13. Relative maximum overtopping volume W_W/h^3 versus the wave overtopping volume parameter E_1 and E_2 with and without relative wave amplitude, respectively, as well as with 10% deviation (dashed line): (**a**) E_1 according to Equation (5) ($R^2 = 0.80$) and (**b**) E_2 according to Equation (7) ($R^2 = 0.80$).

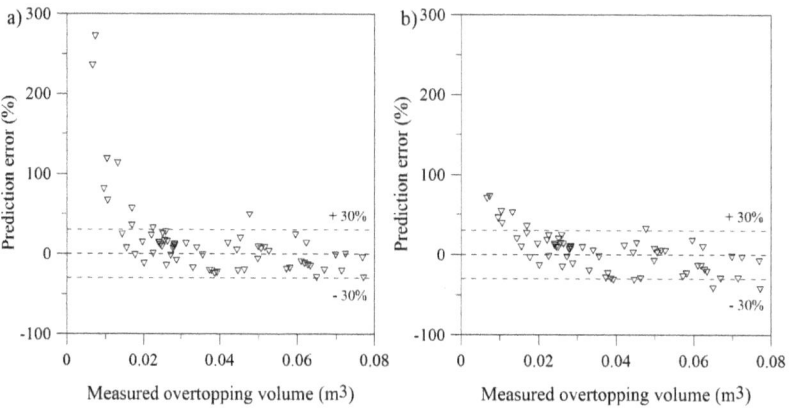

Figure 14. Prediction error for the measured overtopping volume: (**a**) predicted with Equation (4); (**b**) predicted with Equation (6).

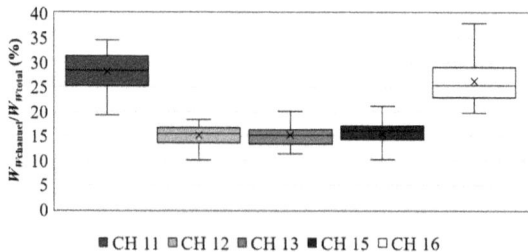

Figure 15. Overtopping volume distribution over the five dam crest sections (see Figure 6). The width of each channel is one fifth of the total dam crest length l_c.

5. Discussion

5.1. Comparison with Other Studies

Other studies include research dealing with dam overtopping due to impulse waves. For example, Müller [19] performed a series of 2D experiments considering the effect of the dam slope and crest width. He proposed an empirical equation for the overtopping volume per unit width, based on the run-up height. Recently Kobel et al. [18] studied dam overtopping due to solitary waves in a 2D laboratory scale set-up, relating to the case in Figure 3b, and presented an empirical equation for predicting the overtopping volume as

$$\frac{W_w}{b_b h^2} = 1.35 \left[\varepsilon \left(\frac{h}{H_d}\right)^{(2/\varepsilon)(\beta/90^0)^{0.25}} \left(\frac{a_w}{b_c}\right)^{0.12} \right]^{0.7} \tag{8}$$

where a_w is the effective wave amplitude defined as $a_w = h + a - H_d$.

It is of interest to compare results from Equation (8) to the measurements of the present study and Equations (4) and (6). Here, two approaches are considered. Approach 1 considers the largest wave amplitude recorded at wave gauge channels closest to the landslide impingement zone (CH 1, CH 2 and CH 3) and assumes that this is a solitary wave. Conversely, Approach 2 considers the wave amplitude recorded at the wave gauges closest to the dam (CH 7, CH 8 and CH 9) and calculates the overtopping volume of a single wave overtopping the dam crest for three sections, the dam inner flank and the outer flanks on each side.

Approach 1 considers that theoretically, the height of a solitary wave (and wave amplitude) as used by Kobel et al. [18] does not decrease and the 2D wave may propagate over unlimited distances without any change of shape. Hence, the parameters of this study are inserted into Equation (8) by assuming that a solitary wave, as in Kobel et al. [18], is generated by the landslide where it impinges the reservoir and that this wave overtops the dam. Thus, the maximum relative wave amplitude recorded in the wave generation zone is used in the calculations. This is a conservative approach, considering the model set-up and the 2D solitary wave type used by Kobel et al. [18]. The ranges of relative amplitudes ε used here by this approach are 0.18 to 0.73 which fall reasonably within the range 0.10 to 0.70 investigated by Kobel et al. [18]. To obtain the overtopping volume, the right-hand side of Equation (8) is multiplied be $b_b h^2$. The mean bottom channel width is used here as b_b.

Approach 2 considers that the waves generated in this study are not 2D solitary waves, but a 3D wave that decays as it propagates towards the dam at different wave propagation angles. Approach 2 provides the overtopping volume at the dam for a single wave. The ranges of ε used in the calculations by this approach are 0.10 to 0.34, which falls within the range 0.10 to 0.70 investigated by Kobel et al. [18] for the use of Equation (8). It should be noted that lower relative wave amplitudes are recorded in some of the tests at CH 7 to 9, but that no overtopping occurs at the respective dam crest sections for $\varepsilon < 0.08$ and $\varepsilon < 0.11$ in case of freeboards of 0.024 and 0.032 m, respectively.

In Figure 16 a comparison is made between predicted and measured overtopping volumes, where the predicted overtopping volume is calculated according to Equations (4) and (6) (deduced for the case in Figure 3a) as well as Equation (8) (deduced for the case in Figure 3b) for Approach 1 and 2. The measured overtopping volume is obtained from the model tests of the present study. Figure 16 demonstrates that as expected, the predictive equation of Kobel et al. [18], using a 2D solitary wave approach (Approach 1) overestimates the experimental data. On average the overestimation is more than twice the measured value. The overestimation factor, represented as the ratio between the predicted and measured overtopping, has a mean value of 2.63 with standard deviation of +/−1.09. In comparison, Equations (4) and (6) extracted from the experimental data, on average, also slightly overestimate the experimental data by a factor of 1.13 and 1.18, respectively, and standard deviations of +/−0.5 and 0.48, respectively.

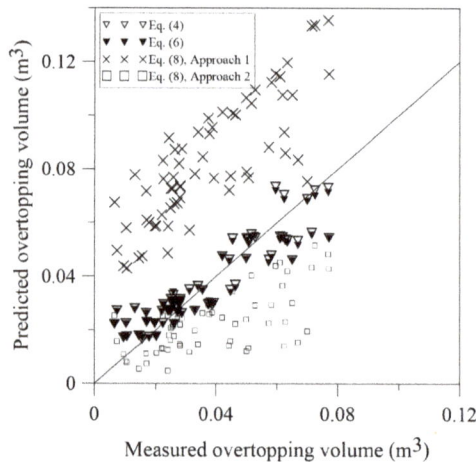

Figure 16. Predicted overtopping volume versus the total overtopping volume measured in the model tests (in m^3) of the present study. The total overtopping volume is predicted by Equations (4) and (6) of the present study and the overtopping volume from a solitary wave using Kobel et al. [18]'s Equation (8) using Approach 1 and 2.

The two Approaches 1 and 2 for the application of Kobel et al. [18]'s Equation (8) are compared in Figure 16. The overtopping volume calculated using Approach 1 overestimates the experimental data while Approach 2 generally underestimates the overtopping volume. For Approach 2, the measured total overtopping volume versus the one predicted by Equation (8), has a mean value of 0.70 with a standard deviation of +/−0.53. The tests that involve large landslide volumes of 4 or 6 blocks, and the highest release height in each case, all result in more than one wave overtopping the dam in the case study. Thus, for these tests the overtopping volume from Approach 2 considering only a single wave largely underestimates the measured total overtopping volume.

5.2. 3D Effects

Figure 16 compares results from studies representing the two cases illustrated in Figure 3, i.e., a 2D study using solitary waves through Approach 1, versus the present 3D study of landslide impinging perpendicular to the reservoir's longitudinal axis and inducing a more complex wave field (Figure 10). Additionally, the 3D wave effects are considered for the 2D formulation with the Approach 2 described above. Figure 16 confirms the statement by Kobel et al. [18] that 2D solitary wave represents the extreme case, considering that the predicted overtopping volume is on average more than twice the experimental results in Approach 1 applying Equation (8). However, Figure 16 also demonstrates that given the relative wave amplitude close to the run-up zone as in Approach 2, the use of Kobel et al. [18]'s, Equation (8) may underestimate the total overtopping volume in cases of narrow reservoirs. Thus, in a risk assessment for the downstream area, threatened by the water overtopping a dam, the water volumes estimated by Equations (4) and (6) are more appropriate for cases as those illustrated in Figures 2a and 3a, but only if these cases can be represented by the physical model of this study (Figures 4 and 6). Approach 1 for the application of Equation (8) by Kobel et al. [18] is appropriate for cases as shown in Figure 3b in case of solitary waves, but is likely to give extreme values, and thus be on the safe side for the cases shown in Figure 3a. Conversely, the 3D Approach 2 to the application of Equation (8) is likely to underestimate the overtopping volume for reservoirs relating to this case study, particularly involving large landslides. Thus, for Approach 2, Equation (8) must be applied cautiously to cases of narrow reservoirs and large landslides because of the potential number of waves overtopping the dam. Combined application of Approaches 1 and 2 for predictions by Equation (8) could be considered, but may give a wide range of potential overtopping volumes.

One important feature extracted from the present study of landslides impinging perpendicular to the longitudinal axis of the reservoir (Figure 3a) is the uneven distribution of the overtopping volume over the dam crest (Figure 15). Equations (4) and (6) provide the total volume of water overtopping the dam. However, from a dam safety perspective the distribution of this along the dam crest is of interest. This distribution can be estimated from Figure 15.

5.3. Limitations and Potential Practical Application

A major limitation of all studies on landslide induced waves overtopping dams is the lack of relevant field data to calibrate physical and numerical models. This is in addition to potential scale effects in physical models, and the simplifications inherent in all modeling, including arising from potential inaccuracies of the instruments. The results from the present study, and other similar ones, should only be used as an aid in risk assessment, assessment of mitigation measures and in decision making that is influenced by dam safety. Furthermore, when applying Equations (4) and (6) the limitations inherent in the fixed model set-up of the present study must be considered. This includes the fixed reservoir geometry, the fixed slope of the landslide ramp, as well as the fixed distance to the landslide ramp. Additionally, the formulation considers only the landslide volume but not the landslide geometry (see Figure 7 and Table 2), but for two slides of the same volume but a different geometry, the wider slide will produce a larger wave and somewhat a larger overtopping volume. Thus, when applying the formulas, the landslide block arrangement (Figure 7) used in the present study must be considered. Furthermore, the slide blocks used represent subaerial rockslides, i.e., those released above the water surface. The results from this study are likely to give extreme values for cases where the landslide extends into the reservoir.

Geological investigations in reservoir areas may identify a potential landslide or rockslide. From the ranges of estimates of the slide size and location in relation to the dam, the equations from this study can be selected for cases that can be represented by Figure 3a and the model set-up (Figures 4 and 6). Assuming the case of Figure 3a, estimation of the necessary parameters to insert into Equation (6) should be attainable, and the same applies to the dam related parameters. However, the estimation must consider the limitations of the fixed reservoir geometry and landslide location stated above. Furthermore, for cases represented by Figure 3a but with very different landslide location from the model set-up (Figures 4 and 6) the results of this study should be used cautiously, with full recognition of the fixed reservoir and landslide set-up. Conversely, the formulation by Kobel et al. [18] requires evaluation of 3D wave parameters (similar to Approach 2) and can then be applied to the more general case, but with caution, at least in the case of a narrow reservoir and a large landslide impinging perpendicular to the reservoir's longitudinal axis. However, the extreme case can be realized with Approach 1 for the use of Kobel et al. [18]'s Equation (8). In any case, the estimated total volume of water overtopping the dam can subsequently be used to realize the potential size of the event threatening the downstream area. However, a time factor is required for calculations of potential flooding of the downstream area (see e.g., Kobel et al. [18]), i.e., resulting in overtopping discharge values. This will be dealt with in a separate article.

The information on the uneven distribution of the overtopping volume may be of importance in assessments for both dam and foundation-related issues, including erosion and scouring. Furthermore, the assessments mentioned can help in decision making regarding the execution of a project in landslide-prone regions, as well as mitigation measures such as monitoring of the landslide, or restricted operation for example through requirements on a minimum freeboard.

6. Conclusions

The research presented deals with the 3D case of a dam overtopped by waves generated as a landslide impinges the reservoir perpendicular to its longitudinal axis (Figure 3a). The governing

J. Mar. Sci. Eng. **2019**, *7*, 221

parameters include the slide volume, landslide release height, landslide speed, slide Froude number, still-water depth, and upstream slope of the dam. In line with the main objective of the paper, an empirical data analysis has been conducted to arrive at Equations (4) and (6) predicting the total volume of water overtopping the dam, based on the fundamental governing variables. The result from the analysis highlights the dominant effect of the slide volume followed by the still-water depth and landslide release height whereas the effect of the upstream dam slope is small. Limitations for the trend equations are stated enabling rough predictions, also in cases where information on some of the parameters may be absent. Furthermore, limitations on the applicability relating to the fixed geometry of the set-up are highlighted.

In line with one of the main objectives of the paper, the experimental data was compared to results using prediction formula derived for the 2D case by Kobel et al. [18]. The comparison confirmed the statement by Kobel et al. [18] that their approach with a 2D solitary wave represents the extreme case, which for the case considered here through Approach 1, resulted in an average overestimation of the measured value by more than a factor of two. However, if the 3D decay of a single wave is considered, as in Approach 2, the prediction by Kobel et al. [18] 's equation generally underestimates the experimental data and may result in an undesirable underestimation in cases where the wave is generated by a large rock slide into a narrow reservoir. In these cases, it is important to estimate the potential number of waves overtopping the dam.

Equations (4) and (6) can, with consideration of the stated limitations, be used for predicting the total volume of water that will overtop a specific dam due to landslide-generated waves with given landslide and dam properties. However, in such an estimation it is of high importance to note the limitation of Equations (4) and (6) arising from the fixed model set-up, i.e., geometry of the basin, location of the landslide ramp, and that the landslide geometry is not considered. Still, the equations can then be used to realize roughly the size of the event threatening the downstream area due to potential landslide-generated waves. The results should only be used to support risk assessment, assessment of mitigation measures, and decision making that is influenced by dam safety.

The overtopping process was described by figures and a quantitative data analysis. One important feature extracted regarding the overtopping process for the 3D case is the variation in time and space (Figure 12) resulting in uneven distributions of the volumes of water overtopping the dam crest (Figure 15). This observation was made possible by the 3D model set-up and is of value for dam safety considerations as well as for foundation-related issues, including erosion and scouring.

Author Contributions: Conceptualization, N.N.T., F.G.S. and L.L.; Formal analysis, N.N.T.; Investigation, N.N.T.; Project administration and Supervision, F.G.S.; Writing, N.N.T.; Writing, review and editing, F.G.S.; Review and editing L.L. and A.K.J.

Funding: Norwegian Water Resources and Energy Directorate (NVE) Project 80051.

Acknowledgments: The support and cooperation of the Norwegian Water Resources and Energy Directorate (NVE) and Grethe Holm Midttømme in the associated study and research program at NTNU is gratefully acknowledged. The contribution of Kiflom W. Belete in the early model development and to technical issues is appreciated. The important contribution of Jochen Aberle in the development of the model, particularly in reducing model effects, is acknowledged. Furthermore, the contribution of Geir Tesaker and Joël Biedermann in the laboratory, and of Ermyas Tamene Haile in the 3D drawing of the model set-up is acknowledged. The first author would like to acknowledge the support from Addis Ababa University, School of Civil and Environmental Engineering, as well as from NTNU, Department of Civil and Environmental Engineering.

Conflicts of Interest: The authors declare no conflict of interest.

Notation

a (m)	wave amplitude;
a_w (m)	effective wave amplitude;
b (m)	slide width;
b_b (m)	reservoir width;
b_c (m)	dam crest width;
E (-)	overtopping volume parameter;
F (-)	Froude number;
f (m)	freeboard;
g (m/s^2)	gravitational acceleration;
h (m)	still-water depth;
h_o (m)	landslide release height;
H (m)	wave height;
H_d (m)	dam height;
l_b (m)	reservoir length;
l_c (m)	crest length;
l_s (m)	slide length;
M (-)	relative slide mass;
m_s (kg)	slide mass;
s (m)	slide thickness;
S (-)	relative slide thickness;
T (-)	wave type product;
v_s (m/s)	landslide speed;
W_S (m^3)	slide volume;
W_W (m^3)	overtopping volume;
α ($^\circ$)	slide ramp angle;
β ($^\circ$)	dam front face angle;
ε (-)	relative wave amplitude;
ρ_w (kg/m^3)	density of water;

References

1. Steven, N.W.; Simon, D. The 1963 landslide and flood at Vaiont reservoir Italy. A tsunami ball simulation. *Ital. J. Geosci.* **2011**, *130*, 16–26.
2. Heller, V.; Hager, W.H.; Minor, H.-E. *Landslide Generated Impulse Waves in Reservoirs: Basics and Computation*; VAW-Mitteilung 211; ETH: Zürich, Switzerland, 2009.
3. Wiegel, R.L. Laboratory studies of gravity waves generated by the movement of a submerged body. *Trans. Am. Geophys. Union* **1955**, *36*, 759–774. [CrossRef]
4. Kamphuis, J.W.; Bowering, R.J. Impulse waves generated by landslides. *Coast. Eng. Proc.* **1970**, *1*, 35. [CrossRef]
5. Watts, P. Water Waves Generated by Underwater Landslides. Ph.D. Thesis, California Institute of Technology, Pasadena, CA, USA, 1997.
6. Synolakis, C. Generation of long waves in laboratory. *J. Waterw. Port Coast. Ocean Eng.* **1990**, *116*, 252–266. [CrossRef]
7. Ataie-Ashtiani, B.; Nik-Khah, A. Impulsive waves caused by subaerial landslides. *Environ. Fluid Mech.* **2008**, *8*, 263–280. [CrossRef]
8. Briggs, M.J.; Synolakis, C.E.; Harkins, G.S.; Green, D.R. Laboratory experiments of tsunami runup on a circular island. *Pure Appl. Geophys.* **1995**, *144*, 569–593. [CrossRef]
9. Fritz, H.M.; Mohammed, F.; Yoo, J. Lituya Bay landslide impact generated mega-tsunami 50th anniversary. In *Tsunami Science Four Years after the 2004 Indian Ocean Tsunami: Part II: Observation and Data Analysis*; Cummins, P.R., Satake, K., Kong, L.S.L., Eds.; Pageoph Topical Volumes; Birkhäuser: Basel, Switzerland, 2009; pp. 153–175.
10. Di Risio, M.; Sammarco, P. Analytical modeling of landslide-generated waves. *J. Waterw. Port Coast. Ocean Eng.* **2008**, *134*, 53–60. [CrossRef]

11. Heller, V.; Spinneken, J. On the effect of the water body geometry on landslide–tsunamis: Physical insight from laboratory tests and 2D to 3D wave parameter transformation. *Coast. Eng.* **2015**, *104*, 113–134. [CrossRef]

12. Romano, A.; Di Risio, M.; Bellotti, G.; Molfetta, M.G.; Damiani, L.; De Girolamo, P. Tsunamis generated by landslides at the coast of conical islands: Experimental benchmark dataset for mathematical model validation. *Landslides* **2016**, *13*, 1379–1393. [CrossRef]

13. McFall, B.C.; Fritz, H.M. Physical modelling of tsunamis generated by three-dimensional deformable granular landslides on planar and conical island slopes. *Proc. R. Soc. A Math. Phys. Eng. Sci.* **2016**, *472*, 20160052. [CrossRef] [PubMed]

14. Bellotti, G.; Romano, A. Wavenumber-frequency analysis of landslide-generated tsunamis at a conical island. Part II: EOF and modal analysis. *Coast. Eng.* **2017**, *128*, 84–91. [CrossRef]

15. Di Risio, M.; Bellotti, G.; Panizzo, A.; De Girolamo, P. Three-dimensional experiments on landslide generated waves at a sloping coast. *Coast. Eng.* **2009**, *56*, 659–671. [CrossRef]

16. Evers, F. *Spatial Propagation of Landslide Generated Impulse Waves*; VAW-Mitteilung 244; ETH: Zürich, Switzerland, 2017.

17. Huber, L.E.; Evers, F.M.; Hager, W.H. Solitary wave overtopping at granular dams. *J. Hydraul. Res.* **2017**, *55*, 799–812. [CrossRef]

18. Kobel, J.; Evers, F.M.; Hager, W.H. Impulse wave overtopping at rigid dam structures. *J. Hydraul. Eng.* **2017**, *143*, 04017002. [CrossRef]

19. Müller, D.R. *Auflaufen und Überschwappen von Impulswellen an Talsperren*; VAW-Mitteilung 137; ETH: Zürich, Switzerland, 1995.

20. Lindstrøm, E.K.; Pedersen, G.K.; Jensen, A.; Glimsdal, S. Experiments on slide generated waves in a 1:500 scale fjord model. *Coast. Eng.* **2014**, *92*, 12–23. [CrossRef]

21. Fritz, H.M.; Hager, W.H.; Minor, H.-E. Landslide generated impulse waves. 1. Instantaneous flow fields. *Exp. Fluids* **2003**, *35*, 505–519. [CrossRef]

22. Fritz, H.M.; Hager, W.H.; Minor, H.-E. Landslide generated impulse waves. 2. Hydrodynamic impact craters. *Exp. Fluids* **2003**, *35*, 520–532. [CrossRef]

23. Mohammed, F.; Fritz, H.M. Physical modeling of tsunamis generated by three-dimensional deformable granular landslides: Landslide generated tsunamis. *J. Geophys. Res. Oceans* **2012**, *117*, C11015. [CrossRef]

24. Zweifel, A. *Impulswellen: Effekte der Rutschdichte und der Wassertiefe*; VAW-Mitteilung 186; ETH: Zürich, Switzerland, 2010.

25. Heller, V.; Spinneken, J. Improved landslide-tsunami prediction: Effects of block model parameters and slide model. *J. Geophys. Res. Oceans* **2013**, *118*, 1489–1507. [CrossRef]

26. Sælevik, G.; Jensen, A.; Pedersen, G. Experimental investigation of impact generated tsunami; related to a potential rock slide, Western Norway. *Coast. Eng.* **2009**, *56*, 897–906. [CrossRef]

27. Heller, V.; Hager, W.H. Wave types of landslide generated impulse waves. *Ocean Eng.* **2011**, *38*, 630–640. [CrossRef]

28. Noda, E. Water waves generated by landslides. *J. Waterw. Harb. Coast. Eng. Div.* **1970**, *96*, 835–855.

© 2019 by the authors. Licensee MDPI, Basel, Switzerland. This article is an open access article distributed under the terms and conditions of the Creative Commons Attribution (CC BY) license (http://creativecommons.org/licenses/by/4.0/).

Journal of
Marine Science and Engineering

MDPI

Article

Capturing Physical Dispersion Using a Nonlinear Shallow Water Model

Rozita Kian [1,*], Juan Horrillo [1], Andrey Zaytsev [2,3] and Ahmet Cevdet Yalciner [4]

[1] Department of Ocean Engineering, Texas A&M University at Galveston, Galveston, TX 77554, USA;
     horrillj@tamug.edu
[2] Laboratory of Computing Hydromechanics and Oceanography, Special Research Bureau for Automation of
     Marine Researches, Far Eastern Branch of Russian Academy of Sciences, Yuzhno-Sakhalinsk 693023, Russia;
     aizaytsev@mail.ru
[3] Department of Applied Mathematics, Nizhny Novgorod State Technical University n.a. R.E. Alekseeva,
     Nizhny Novgorod 603950, Russia
[4] Department of Civil Engineering, Ocean Engineering Research Center, Ankara 06800, Turkey;
     yalciner@metu.edu.tr
*   Correspondence: rozita@tamu.edu; Tel.: +1-409-740-4589

Received: 9 May 2018; Accepted: 4 July 2018; Published: 9 July 2018

Abstract: Predicting the arrival time of natural hazards such as tsunamis is of very high importance to the coastal community. One of the most effective techniques to predict tsunami propagation and arrival time is the utilization of numerical solutions. Numerical approaches of Nonlinear Shallow Water Equations (NLSWEs) and nonlinear Boussinesq-Type Equations (BTEs) are two of the most common numerical techniques for tsunami modeling and evaluation. BTEs use implicit schemes to achieve more accurate results compromising computational time, while NLSWEs are sometimes preferred due to their computational efficiency. Nonetheless, the term accounting for physical dispersion is not inherited in NLSWEs, calling for their consideration and evaluation. In the present study, the tsunami numerical model NAMI DANCE, which utilizes NLSWEs, is applied to previously reported problems in the literature using different grid sizes to investigate dispersion effects. Following certain conditions for grid size, time step and water depth, the simulation results show a fairly good agreement with the available models showing the capability of NAMI DANCE to capture small physical dispersion. It is confirmed that the current model is an acceptable alternative for BTEs when small dispersion effects are considered.

Keywords: Nonlinear Shallow Water Equations; NAMI DANCE model; Boussinesq-Type Equations; grid size

1. Introduction

Tsunami is a very common hazard that often devastates coastal areas. The prediction of tsunami arrival time is crucial for the effective and wise decisions by tsunami warning centers. As a result, numerical studies have been extensively carried out to simulate tsunami propagation more effectively and efficiently. Traditionally, models utilizing NLSWEs have been used to simulate tsunami generation and to predict tsunami propagation and runup. Notwithstanding the great advantage of very low computational costs, NLSWEs based models do not consider physical dispersion. Alternatively, the utilization of dispersive models using the Boussinesq approach is rapidly growing in tsunami modeling. Dispersive models are also used for the calculation of tsunami propagation in the near shore regions. Nevertheless, this type of models are less computationally efficient as compared to NLSWEs based approaches.

The coastal modeling community has put a lot of attention on modeling wave propagation using BTEs models and, accordingly, many improvements have been reported for the tsunami generation, propagation and runup modeling. Three major advantages of using BTEs over NLSWEs are: the horizontal particle velocity and acceleration over the depth are considered to be varying and the vertical acceleration is not neglected; better prediction of wave propagation over complex topographies can be achieved; and better frequency dispersion effects can be obtained [1]. Although Navier-Stokes (NS) models are fully dispersive due to the consideration of the complete physics, it is impractical to perform a full solution of the NS equation over large domains, especially for tsunami propagation. Notwithstanding the capability of NS and BTEs models, Horrillo et al. [2] showed that for practical purposes solutions that are based on the NLSWEs are quite reliable in many circumstances since they give consistent results compared to BTEs and full NS models. More importantly, as discussed, models based on NLSWEs have the great advantage of very low computational costs. Further improvements of NLSWEs models can be achieved by taking the advantage of inherent numerical dispersion errors mimicking physical frequency dispersion of tsunamis. Generally, frequency dispersion consideration is very important to get a precise and realistic prediction of tsunami propagation. In fact, many factors such as refraction, diffraction, nonlinear shoaling, wave-wave interaction, breaking and runup heavily depend on dispersion effects and should be considered to obtain an accurate prediction.

Linear shallow water equations considering numerical dispersion were solved by Imamura et al. [3] for principal horizontal axes. Cho [4] improved the algorithm to include the effects of frequency dispersion for diagonal propagation. They both used an algorithm through the modified leap-frog finite difference scheme for discretizing linear shallow water equations to account for the effect of frequency dispersion. This helped to control the inherent numerical dispersion errors of the scheme to cover the effects of physical frequency dispersion for the propagation of tsunami in bathymetries with constant water depth. Later, the approach was extended by Yoon [5] to a slowly varying bathymetry. He proposed a hidden grid system in which the local water depth determines the grid size for the capturing of dispersion effects. Subsequently, Yoon et al. [6] developed another scheme using an explicit finite difference method to solve the linearized BTEs and to improve the physical frequency dispersion effects. The linearized BTEs model considers the frequency dispersion effects from shallow to intermediate water regimes accurately [7–14]. However, it needs finer spatial and temporal resolutions since the higher order of the derivatives associated with the consideration of the frequency dispersion requires more computational efforts, which is sometimes unnecessary. Establishing an appropriate grid size in NLSWEs models in which the numerical error matches physical frequency dispersion may lead to similar results to BTEs based models with less expensive computational costs.

One of the most commonly used tsunami numerical models is Tohoku University's Numerical Analysis Model for Investigation of Near-field tsunamis, No. 2 (TUNAMI-N2) [15] which solves NLSWEs using the finite difference method. This model uses the explicit leap-frog scheme and has been modified to account for frequency dispersion [3]. The dispersion effects are captured by choosing an appropriate time step and a grid size given by [3]:

$$\Delta x = \sqrt{4h^2 + gh(\Delta t)^2} \tag{1}$$

where Δx is the grid size, h denotes water depth, g symbolizes gravity acceleration and Δt is the time step.

Using a similar approach accounting for the frequency dispersion effects and by using Equation (1) following the efforts of Imamura et al. [3] and Cho [4], a shallow water equations based model named NAMI DANCE was developed [16]. A brief explanation of NAMI DANCE is presented in the next section. In this study, the ability of NAMI DANCE to capture dispersion effects is examined with several bathymetry profiles or case studies taken from the literature [5,6]. The bathymetry profiles are: (i) uniform bottom, (ii) slowly varying depth and (iii) submerged circular shoal. Our objective is to investigate the fact that by selecting an appropriate grid size in NAMI DANCE, we can obtain comparable results to BTEs or NS models in mimicking physical dispersion effects.

2. NAMI DANCE Tool

NAMI DANCE, tsunami numerical modeling was developed by the Special Bureau of Automation of Research Russian Academy of Sciences (SBARRAS) in collaboration with the Middle East Technical University (METU) in Turkey. The code uses the leap-frog finite difference scheme of Goto et al. [17] and is capable of solving NLSWEs with various boundary conditions [16,18,19]. The model has the capability to switch off the nonlinearity and allows nested domains in Cartesian or spherical coordinate systems. NAMI DANCE has been verified and validated with standard benchmark problems as well as with several tsunami events worldwide [18–26]. The governing equations of NAMI DANCE are the nonlinear form of shallow water equations and hence, the vertical motions of water particles do not have any role on the hydrostatic pressure. More information about NAMI DANCE is available in Ref. [16].

3. Gaussian Wave Propagation over Uniform Bottom

The accuracy of the numerical solution of NLSWEs in NAMI DANCE is tested through simulation of a Gaussian hump tsunami as initial free surface [5,6] on a constant uniform bottom. In this section, the water surface displacement at certain time steps along horizontal, vertical and diagonal directions are computed and compared with the analytical solution of the linear BTEs [3,27]. The initial form of the water surface is determined using Gaussian function Equations (2) and (3) resulting in the cross sectional shape given in Figure 1.

$$\eta(r, \theta) = e^{-(r/a)^2} \tag{2}$$

$$\frac{\partial \eta(r, \theta)}{\partial t} = 0 \tag{3}$$

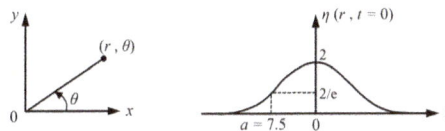

Figure 1. Profile of initial free surface and the coordinate system (adopted from [5]).

In Equation (2) a denotes the characteristic radius of Gaussian function, $r = (x^2 + y^2)^{1/2}$ is the distance of the Gaussian hump from the center, θ represents the angle measured counter clockwise from the x axis, t represents time and η is free surface elevation. Shown in Figure 2 are the plan view of the initial wave condition, propagation axes, and location of gauges. The analytical solution of the initial Gaussian wave profile was solved by the linear form of the Boussinesq equation given by Carrier [27] (see Figure 3):

$$\eta(r, t) = \int_0^\infty a^2 e^{-(a/k)^2/4} k \cos\left(\frac{\sqrt{gh}\, kt}{\sqrt{1 + \frac{(kh)^2}{3}}}\right) J_0(kr) dk \tag{4}$$

where J_0 is the zero order Bessel function of the first kind, k denotes the wave number of uniform waves, $g = 9.81$ m/s^2, a is the characteristic radius and h represents the water depth.

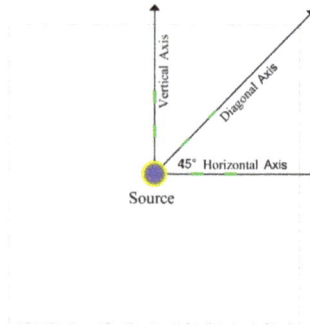

Figure 2. Plan view of the initial wave and axis for the uniform bottom domain. Green lines represent gauges along the axis.

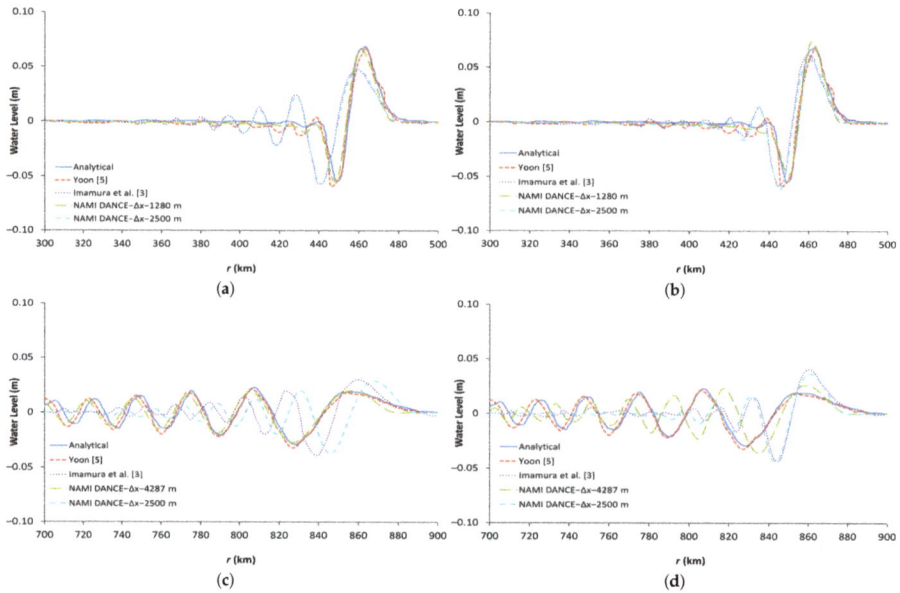

Figure 3. Comparison of water surface profiles of several numerical methods against the analytical solution of Carrier [27]: (**a**) $h = 600$ m and $\theta = 0°, 90°$, (**b**) $h = 600$ m and $\theta = 45°$, (**c**) $h = 2100$ m and $\theta = 0°, 90°$, (**d**) $h = 2100$ m and $\theta = 45°$. All comparisons are done at time 600 s for two different propagation axes and depths.

Figure 3 compares snapshots of the water surface profiles at $t = 600$ seconds using different numerical methods against the analytical solution of Carrier [27]. Two different propagation axes ($\theta = 0°$ and $\theta = 45°$) as well as two different depths ($h = 600$ m and $h = 2100$ m) are investigated using a uniform bottom domain. The time step is not strongly influenced by a ratio for dispersion and hence, a constant value of $\Delta t = 6$ s is used following the work of Yoon [5]. The grid size Δx is constant according to Equation (1) and the memory capacity of the hardware/software. In the NAMI DANCE simulations, two different methods are employed to specify the grid size. In the first case, the grid size is selected according to Equation (1) which provides $\Delta x = 1280$ m, given $h = 600$ m (Figure 3a,b) and $\Delta x = 4287$ m for $h = 2100$ m (Figure 3c,d). In the second case, the grid size is selected as $\Delta x = 2500$ m according to [5]. Figure 3a,c ($\theta = 0°$) show that when Equation (1) is satisfied, NAMI DANCE's solutions are closer to the analytical results of [27] and Yoon's model [5], while, in diagonal propagation (Figure 3b,d), Yoon's [5] solution leads to more accurate results. Along the diagonal propagation axis ($\theta = 45°$), more wave dispersion is observed in Imamura et al.'s [3] results since the dispersion correction term in the diagonal direction is not considered in the solutions, whereas the numerical solution of Yoon is independent of direction [5]. When Equation (1) is not satisfied, the numerical solutions of NAMI DANCE do not fit well with the analytical results with the tendency to be closer to Imamura et al.'s [3] results. It should be mentioned that Yoon's model considered the dispersion correction term in the diagonal direction [5]. Yoon's model is independent of the direction, and therefore its result at $\theta = 45°$ is closer to the analytical solution as compared to those of Imamura's [3].

Figure 3c,d show that when Equation (1) is satisfied (for $\Delta x = 4287$ m) both Yoon's [5] and NAMI DANCE's results are closer to the analytical solutions compared to other models with Yoon's solution [5] being more accurate in the diagonal propagation. In general, the NAMI DANCE's results for $h = 600$ m, $\Delta x = 1280$ m and $h = 2100$ m, $\Delta x = 4287$ m compare well with the analytical solution (see Figure 3). Figure 3 also indicates that whenever the Equation (1) is satisfied, physical dispersion of the analytical solution is reproduced with reasonable accuracy by the numerical dispersion of the leap-frog finite difference scheme of NAMI DANCE.

In a separate study, Yoon et al. in 2007 [6] improved the numerical scheme to reproduce the physical dispersion effects and tested on a uniform bottom domain with different water depths [6]. Shown in Figure 4 are the results which compare time histories of the water surface profiles at a distance of 312,900 m ($150\Delta x$ and $\Delta x = 2086$ m) away from the center of the initial Gaussian hump obtained by different models including Yoon et al.'s [6] model. Here, three different depths and two different propagation axes are investigated ($h = 500$ m, $h = 1000$ m and $h = 1500$ m with $\theta = 0°$ and $\theta = 45°$). Again, in the simulations with the NAMI DANCE model two different methods are employed to specify the grid size. In the first case, the grid size is selected according to Equation (1) which provides $\Delta x = 1085$ m for $h = 500$ m (Figure 4a,b), $\Delta x = 2086$ m for $h = 1000$ m (Figure 4c,d) and $\Delta x = 3087$ m for $h = 1500$ m (Figure 4e,f). In the second case, the grid size is selected as $\Delta x = 2086$ m according to [6]. Figure 4a,c,e ($\theta = 0°$) show that when Equation (1) is satisfied, NAMI DANCE's results are closer to the analytical solution of Carrier [27] and Yoon et al.'s [6] results as compared to other models, while, in diagonal propagation (Figure 4b,d,f) Yoon et al.'s [6] solution leads to more accurate results. On the other hand, more dispersion effects are observed in Imamura et al.'s [3] results, since the dispersion correction term in the diagonal direction is not considered in the solutions. When Equation (1) is not satisfied in NAMI DANCE, the results do not fit well with the analytical solutions having a tendency to be closer to Imamura et al.'s [3] results. In addition, on the diagonal propagation, NAMI DANCE results show less dispersive behavior as compared to other models. The solutions predicted by Yoon et al.'s [6] model are closer to the analytical solution than those of Imamura et al.'s [3] model since the dispersion correction term in the diagonal direction is considered in Yoon et al.'s [6] model.

In general, NAMI DANCE's results for $\Delta x = 1085$ m given $h = 500$ m, $\Delta x = 2086$ m given $h = 1000$ m and $\Delta x = 3087$ m given $h = 1500$m, compare well with the analytical solution. However, along the diagonal propagation axis the water surface profiles are slightly higher than the analytical solution. In addition, whenever Equation (1) is satisfied in NAMI DANCE, physical dispersion of the Boussinesq equations is reproduced with reasonable accuracy by the numerical dispersion of the leap-frog finite difference scheme. It is also worth mentioning that the choice of the space discretization by Equation (1) can unresolve the wave length, given the insufficient points per wave length to provide the approximation needed for accuracy of numerical solutions, as pointed out in earlier work [3]. However, this basic low resolution (points per wave length) is needed to create the numerical dispersion required to reproduce the physical dispersion.

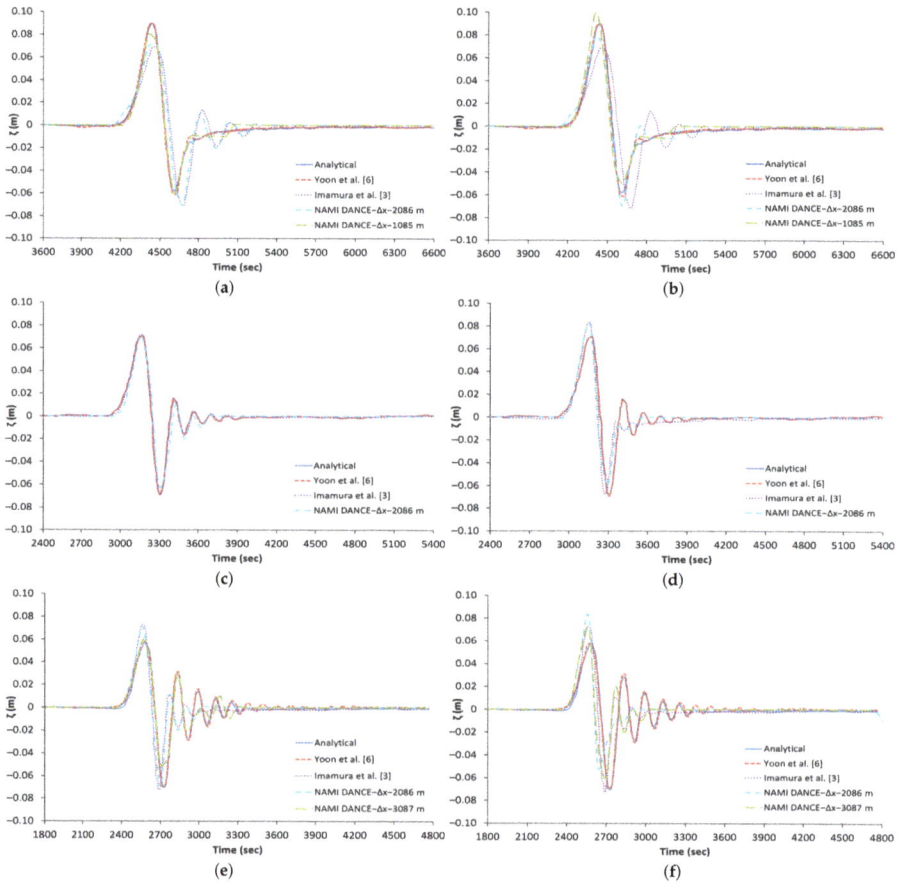

Figure 4. Comparison of water surface profiles of several numerical methods against the analytical solution of Carrier [27]: (**a**) $h = 500$ m and $\theta = 0°, 90°$, (**b**) $h = 500$ m and $\theta = 45°$, (**c**) $h = 1000$ m and $\theta = 0°, 90°$, (**d**) $h = 1000$ m and $\theta = 45°$, (**e**) $h = 1500$ m and $\theta = 0°, 90°$, (**f**) $h = 1500$ m and $\theta = 45°$. All comparisons are done at distance 312,900 m away from the center of the initial Gaussian hump on a uniform bottom.

4. Gaussian Wave Propagation over Slowly Varying Depth

The above-mentioned wave propagation model (proposed by Yoon [5]) uses a uniform bottom and therefore is not applicable for a varying bathymetry. Alternatively, Yoon [5] developed a new finite difference adjustable grid scheme which satisfies the local wave frequency dispersion in both varying and flat bathymetries. He showed that this new scheme provides a noticeable improvement on the effect of frequency dispersion compared to the schemes of Imamura et al. [3] and Cho [4].

In this section, the one-dimensional water surface displacement at certain time along the horizontal direction is computed with NAMI DANCE and compared with the solutions of Yoon's [5] model, adjustable grid model [5] and Imamura et al.'s [3] model. The wave propagation is over a slowly varying bathymetry consisting of two uniform water depths, $h = 2$ km and 1 km, connected by a half-sine function (Figure 5a). The initial water surface profile is determined using the same Gaussian profile defined by Equations (2) and (3) (Section 3). Figure 5b,c compare snapshots of the water surface profiles at time 40,000 s using different numerical methods indicated above against the numerical solution of Yoon's (2002) adjustable grid. In the simulations, NAMI DANCE employed two different grid sizes selected according to the Equation (1) which provides $\Delta x = 4$ km, given $h = 2$ km (Figure 5b) and $\Delta x = 2$ km for $h = 1$ km (Figure 5c). Simulations were done using a uniform grid size for the entire domain, even though a slowly varying bottom with two different depths were considered.

In general, results show that when Equation (1) is satisfied, NAMI DANCE's solutions are closer to both Yoon's [5] models. However, when a low resolution grid size (4 km) was used in NAMI DANCE, a smaller dispersive effect was observed in the tail of the wave train. Since the large grid size leads to having a smaller numerical dispersion than physical dispersion, smaller tailing waves are created.

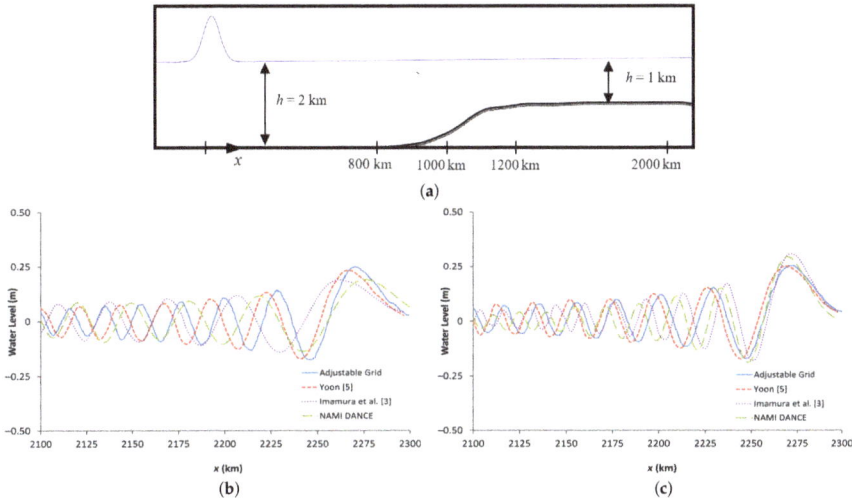

Figure 5. (**a**) Schematic representation of Gaussian hump propagating over a slowly varying topography [5]. Water surface profiles comparison of several methods at $t = 40,000$ s and $h = 1$ km for two different grid sizes (**b**) $\Delta x = 4$ km and (**c**) $\Delta x = 2$ km.

5. Solitary Wave Propagation over a Submerged Circular Shoal

Using a finite difference model, Yoon et al. [6] developed a dispersion-correction parameter to solve Boussinesq type wave equations which could be applied to more general varying bathymetries, e.g., two dimensions. In the model proposed by Yoon et al. [6], the numerical dispersion effect is determined by the local water depth variations and time step, presenting good performance especially for cases featuring short waves. Yoon et al. [6] presented the evolution of a tsunami wave over a submerged circular shoal and since there is no analytical solution available for this case, they compared their solutions with those of the linearized Nwogu's Boussinesq equations [7] implemented in the FUNWAVE code [8,9].

FUNWAVE is a fully nonlinear Boussinesq wave model and has been improved for the dispersion relationship of short waves. In BTEs based models, the grid size should be small enough to avoid the dependency of numerical dispersion effects on higher order terms. Therefore, FUNWAVE's results with smaller grid size are more reliable.

Figure 6 shows the domain configuration with size of 1500 km × 500 km. The center of the submerged circular shoal is located at $x_0 = 500$ km, $y_0 = 250$ km. The water depth of the numerical domain over the shoal is given by:

$$h(r) = \begin{cases} 1500, & r \geq R_1 \\ 1500 \left(\frac{r^2}{R_1^2} \right), & R_2 < r < R_1 \\ 500, & r \leq R_2 \end{cases} \tag{5}$$

where $r = \sqrt{(x-x_0)^2 + (y-y_0)^2}$ is the distance from the center of the submerged circular shoal, $R_1 = 150$ km is the base radius and $R_2 = 86.6$ km is the top radius.

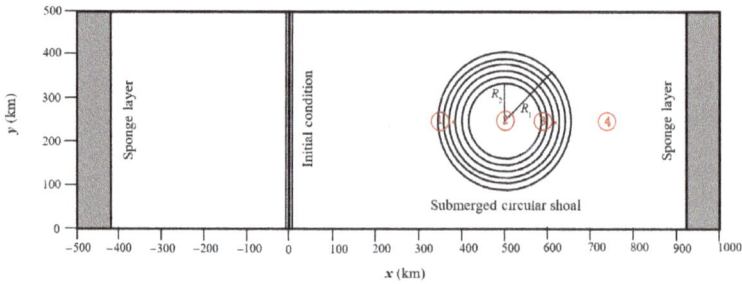

Figure 6. Submerged circular shoal, wave gauges and initial condition [6].

The initial form of the water surface is given in the form of a soliton as:

$$\eta(x) = 2e^{-(x/7500)^2} \tag{6}$$

The gauges 1 and 3 are located on the slope of the circular shoal with depth of 1000 m. Gauge 2 is placed at the center of the circular shoal with 500 m water depth and Gauge 4 is at $x = 750$ km, see Figure 6. In order to dissipate the outgoing wave energy, Yoon et al. [6] considered sponge layers at the two ends of the numerical domain. The grid size was selected as $\Delta x = 2000$ m and the time step $\Delta t = 4$ s. The water surface displacements at these numerical gauges were also computed with NAMI DANCE and compared with the FUNWAVE's results to test our model capability to reproduce the physical dispersion effects. Simulations with FUNWAVE were conducted with two different grid sizes ($\Delta x = 500$ m and $\Delta x = 2000$ m). The grid size in NAMI DANCE is selected as 3000 m to satisfy Equation (1).

Figure 7 compares time series of the water surface profiles at different numerical gauges. The simulations are carried out for 9000 seconds from the initial soliton wave condition. In general, the first wave matches well with the FUNWAVE's results but later NAMI DANCE's results show less dispersive behavior compared to both, Yoon et al. [6] and FUNWAVE [6]. Figure 7a shows the time history of water surface profiles recorded at numerical gauge 1 (1000 m water depth). It is clearly seen that NAMI DANCE matches very well with the other results, especially with FUNWAVE ($\Delta x = 2000$ m), although, some small discrepancies are seen in the tail of the time series. This good match is attributed to the model capability to mimic the physical dispersion and also to the minor effects caused by refraction and shoaling. Figure 7b shows the time history of water surface profiles computed at numerical gauge 2 which is located at the center of the circular shoal at a water depth of 500 m. In general, NAMI DANCE's first two wave results are in good agreement with both, FUNWAVE and Yoon et al.'s [6] results. The first wave is slightly under predicted, as a consequence, a small phase lag is observed in the following wave. When the first wave is smaller than predicted, tail waves occur in order to satisfy the energy conservation.

The time history of the water surface profile on the back side of the circular shoal (1000 m depth) at numerical gauge 3 is shown in Figure 7c. At this gauge, NAMI DANCE's result shows a smaller wave amplitude and a slightly longer period. However, in general, results are considered reasonable if we account the accumulating effect of numerical dispersion through the shoaling process. Figure 7d confirms the model's capability to reproduce well the water surface profile at the lee part of the circular shoal (1500 m depth), numerical gauge 4. Although the wave went through a varying bathymetry as well as the shoaling process, the agreement of the NAMI DANCE solution with FUNWAVE result is quite good indicating that the model approach works well in mimicking the physical dispersion proving that a correct grid size and time step are chosen. Thus, the NAMI DANCE model is a good option to be applied in practical tsunami modeling that requires dispersion effects.

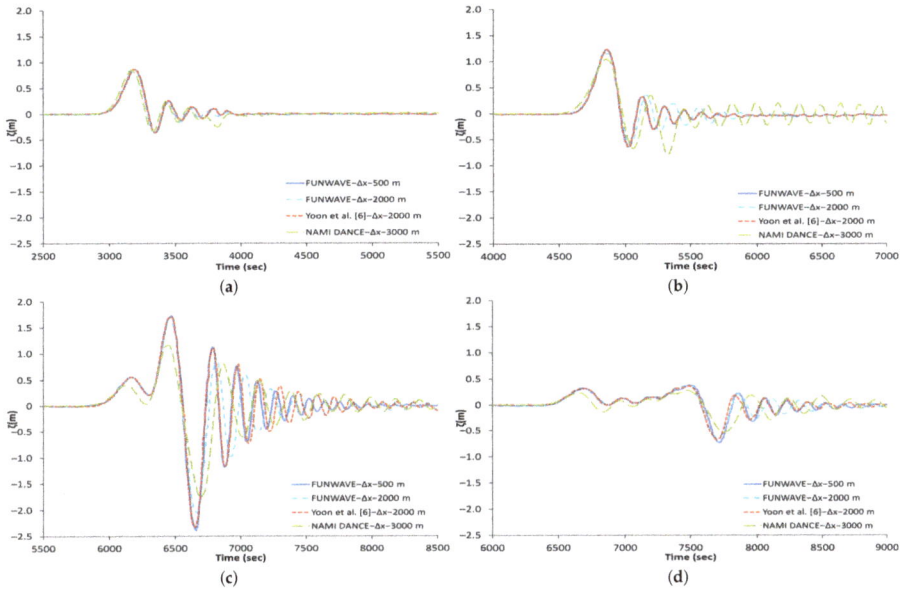

Figure 7. Comparison of time history of water surface profiles of NAMI DANCE, FUNWAVE (using Nwogu's BTEs) [8,9] and Yoon et al.'s [6] solution at numerical gauges: (**a**) Gauge 1, (**b**) Gauge 2, (**c**) Gauge 3 and (**d**) Gauge 4.

It is noticeable that the first waves play the main role in tsunami inundation problems, however, inundation is outside the scope of this study.

6. Conclusions

In this study, NAMI DANCE was tested with several bathymetry profiles and compared with previous works in the literature in order to test the capability of the model for reproducing the numerical dispersion. Based on the cases investigated in the current study, it was found that the model can reproduce reasonable physical frequency dispersion provided that for each case an appropriate grid size is used based on the time step and water depth (Equation (1)). This makes the model an attractive candidate to be used in simplified and practical tsunami calculations since the approach is more efficient as compared with more sophisticated models based on BTEs. It should be emphasized that this is a purely numerical approach, causing the numerical errors to mimic some physics that is not included in the solved equations. Following certain conditions for the grid size based on Equation (1), the simulation results showed a fairly good agreement with the available models, confirming NAMI DANCE capability to capture small physical dispersion. It is also worth mentioning that the choice of the space discretization by Equation (1) can resolve the issue of having not enough points per wave length to provide the approximation needed for accuracy of numerical solutions. This was also pointed out in the earlier work of Imamura et al. [3]. This resolution (points per wave length) is needed to create the numerical dispersion required to capture the physical dispersion. In general, NAMI DANCE follows the solutions based on BTEs as well as analytical models more closely as compared to that of Imamura et al. [3], still more work will be necessary in the near future, especially to account for the more complex bathymetry profiles required in practical application of tsunami.

Author Contributions: Conceptualization, R.K. and A.C.Y.; Methodology, R.K.; Software, A.Z., A.C.Y.; Validation, J.H; Formal Analysis, R.K.; Investigation, R.K.; Resources, R.K.; Data Curation, J.H.; Writing—Original Draft Preparation, R.K.; Writing—Review & Editing, J.H.

Funding: This study is partly supported by The Scientific and Technological Research Council of Turkey (TUBITAK), 2215 Program. It is also partially funded by framework of the state task program in the sphere of scientific activity of the Ministry of Education and Science of Russian Federation (project No. 5.5176.2017/8.9) and Assessment, Strategy and Risk Reduction for Tsunamis in Europe (ASTARTE) project - FP7-ENV2013 6.4-3, Grant 603839.

Acknowledgments: R.K. would like to thank the faculty summer research support from Texas A&M University at Galveston. Also, The authors would like to express their gratitude to the editor and the reviewers for their constructive comments and suggestions that helped enhancing the quality of the present work.

Conflicts of Interest: The authors declare no conflict of interest.

Abbreviations

The following abbreviations are used in this manuscript:

BTE   Boussinesq-Type Equation
NS   Navier-Stokes
NLSWE Nonlinear Shallow Water Equation

References

1. Pujiraharjo, A.; Hosoyamada, T. Comparison of numerical wave models of long distance tsunami propagation—An application to Indian ocean tsunami in 2004. *J. Appl. Mech.* **2007**, *10*, 749–756. [CrossRef]
2. Horrillo, J.; Kowalik, Z.; Shigihara, Y. Wave dispersion study in the Indian ocean-tsunami of December 26, 2004. *Mar. Geod.* **2006**, *29*, 149–166. [CrossRef]
3. Imamura, F.; Shuto, N.; Goto, C. Numerical simulations of the transoceanic propagation of tsunamis. In *Proceedings of the 6th Congress of the Asian and Pacific Regional Division*; IAHR: Kyoto, Japan, 1988.
4. Cho, Y.-S. Numerical Simulations of Tsunami Propagation and Run-Up. Ph.D. Thesis, School of Civil and Environmental Engineering, Cornell University, Ithaca, NY, USA, January 1995.
5. Yoon, S.B. Propagation of distant tsunamis over slowly varying topography. *J. Geophys. Res. Ocean.* **2002**, *107*, 1–11. [CrossRef]
6. Yoon, S.B.; Lim, C.H.; Choi, J. Dispersion-correction finite difference model for simulation of transoceanic tsunamis. *Terr. Atmos. Ocean. Sci.* **2007**, *18*, 31–53. [CrossRef]
7. Nwogu, O. Alternative form of Boussinesq equations for nearshore wave propagation. *J. Waterw. Port Coast. Ocean Eng.* **1993**, *119*, 618–638. [CrossRef]
8. Wei, G.; Kirby, J.T. Time-dependent numerical code for extended Boussinesq equations. *J. Waterw. Port Coast. Ocean Eng.* **1995**, *121*, 251–261. [CrossRef]
9. Kirby, J.T.; Wei, G.; Chen, Q.; Kennedy, A.B.; Dalrymple, R.A. *Fully Nonlinear Boussinesq Wave Model*; User Manual, Rep. No. Technical Report, CACR-98-06; University of Delaware: Newark, DE, USA, 1998.
10. Lynett, P.J.; Liu, P.L.-F. A two-dimensional, depth-integrated model for internal wave propagation over variable bathymetry. *Wave Motion* **2002**, *36*, 221–240. [CrossRef]
11. Hsiao, S.C.; Lynett, P.; Hwung, H.-H.; Liu, P.L.-F. Numerical simulations of nonlinear short waves using a multilayer model. *J. Eng. Mech.* **2005**, *131*, 231–243. [CrossRef]
12. Lynett, P.J. Nearshore wave modeling with high-order Boussinesq-type equations. *J. Waterw. Port Coast. Ocean Eng.* **2006**, *132*, 348–357. [CrossRef]
13. Lynett, P.J. Effect of a shallow water obstruction on long wave runup and overland flow velocity. *J. Waterw. Port Coast. Ocean Eng.* **2007**, *133*, 455–462. [CrossRef]
14. Wang, X. Numerical Modelling of Surface and Internal Waves over Shallow and Intermediate Water. Ph.D. Thesis, Cornell University, Ithaca, NY, USA, 2008.
15. Imamura, F. Review of tsunami simulation with a finite difference method, long-wave runup models. In *Proceedings of the International Workshop, Friday Harbour, USA*; World Scientific: Singapore, 1996; pp. 25–42.

16. Yalciner, A.C.; Pelinovsky, E.; Zaytsev, A.; Kurkin, A.; Ozer, C.; Karakus, H. *Nami Dance Manual*; Middle East Technical University, Civil Engineering Department, Ocean Engineering Research Center: Ankara, Turkey, 2006. Available online: http://namidance.ce.metu.edu.tr/pdf/NAMIDANCE-version-5-9-manual.pdf (accessed on 10 June 2018).

17. Goto, C.; Ogawa, Y.; Shuto, N.; Imamura, F. *IUGG/IOC Time Project: Numerical Method of Tsunami Simulation with the Leap-Frog Scheme*; Manuals and Guides; Intergovernmental Oceanographic Commission of UNESCO: Paris, France, 1997; Volume 35.

18. Kian, R.; Yalciner, A.C.; Zaytsev, A. Dispersion of long waves on varying bathymetry. *Coast. Eng. Proc.* **2014**, *1*, 19. [CrossRef]

19. Kian, R.; Yalciner, A.C.; Zaytsev, A. Evaluating the performance of tsunami propagation models. In Proceedings of the Forecast Engineering, Bauhaus Summer School, Weimar, Germany, 17–29 August 2014; Volume 25, p. 10.

20. Velioglu, D.; Kian, R.; Yalciner, A.C.; Zaytsev, A. Performance assessment of NAMI DANCE in tsunami evolution and currents using a benchmark problem. *J. Mar. Sci. Eng.* **2016**, *4*, 49. [CrossRef]

21. Kian, R.; Velioglu, D.; Yalciner, A.C.; Zaytsev, A. Effects of harbor shape on the induced sedimentation; L-type basin. *J. Mar. Sci. Eng.* **2016**, *4*, 55. [CrossRef]

22. Ozer, C. Tsunami Hydrodynamics in Coastal Zones. Ph.D. Thesis, Middle East Technical University, Ankara, Turkey, 2007.

23. Ozer, C.; Yalciner, A.C. Sensitivity study of hydrodynamic parameters during numerical simulations of tsunami inundation. *Pure Appl. Geophys.* **2011**, *168*, 2083–2095. [CrossRef]

24. Yalciner, A.C.; Ozer, C.; Karakus, H.; Zaytsev, A.; Guler, I. Evaluation of coastal risk at selected sites against eastern Mediterranean tsunamis. *Coast. Eng. Proc.* **2011**, *1*, 10. [CrossRef]

25. Yalciner, A.C.; Suppasri, A.; Mas, E.; Kalligeris, N.; Necmioglu, O.; Imamura, F.; Ozer, C.; Zaytsev, A.; Ozel, M.N.; Synolakis, C. Field survey on the coastal impacts of march 11, 2011 great east Japan tsunami. In Proceedings of the Seismic Protection of Cultural Heritage, Antalya, Turkey, 31 October–1 November 2011.

26. Yalciner, A.C.; Zaytsev, A.; Aytore, B.; Insel, I.; Heidarzadeh, M.; Kian, R.; Imamura, F. A possible submarine landslide and associated tsunami at the northwest Nile delta, Mediterranean Sea. *Oceanography* **2014**, *27*, 68–75. [CrossRef]

27. Carrier, G.F. Tsunami propagation from a finite source. In Proceedings of the 2nd UJNR Tsunami Workshop, NGDC, Honolulu, HI, USA, 5–6 November 1990.

© 2018 by the authors. Licensee MDPI, Basel, Switzerland. This article is an open access article distributed under the terms and conditions of the Creative Commons Attribution (CC BY) license (http://creativecommons.org/licenses/by/4.0/).

Journal of
Marine Science and Engineering

MDPI

Article

The historical reconstruction of the 1755 earthquake and tsunami in downtown Lisbon, Portugal

Angela Santos [1,*], Mariana Correia [1], Carlos Loureiro [2], Paulo Fernandes [2,3] and Nuno Marques da Costa [1]

[1] Centre for Geographical Studies, Institute of Geography and Spatial Planning, Universidade de Lisboa, Rua Branca Edmée Marques, 1600-276 Lisboa, Portugal
[2] Museu de Lisboa, EGEAC, Campo Grande 245, 1700-091 Lisboa, Portugal
[3] Centro de Estudos em Arqueologia, Artes e Ciências do Património, Universidade de Coimbra, 3000-395 Coimbra, Portugal
* Correspondence: angela.santos@campus.ul.pt; Tel.: +351-210-443-000

Received: 1 April 2019; Accepted: 27 June 2019; Published: 4 July 2019

Abstract: The historical accounts of the 1755 earthquake and tsunami in Lisbon are quite vast providing a general overview of the disaster in the city. However, the details remain unknown. Therefore, the objective of this research is to understand and reconstruct the impact of the 1755 event (earthquake, tsunami, and fire) in downtown Lisbon. Thus, the historical data has been compiled and analyzed, to complement tsunami modeling and a field survey. Although census data are not very accurate, before the disaster there were about 5500 buildings and about 26,200 residents in downtown Lisbon; after the disaster, no records of the buildings were found and there were about 6000–8800 residents. There were about 1000 deaths in the study area. The results also show that the earthquake did not cause significant damage to most of the study area, which contradicts general knowledge. After the earthquake, a fire started that quickly spread throughout the city causing most damage to property. The tsunami hit mostly the west and central parts of the study area. The numerical model results show the tsunami hit the studied area about 60 min after the earthquake, inundating the seafront streets and squares up to 200 m inland. In addition, two major waves were calculated, which are in agreement with the historical accounts.

Keywords: 1755 tsunami; downtown Lisbon; historical data; numerical model

1. Introduction

The 1 November 1755 earthquake triggered a tsunami that hit the entire Portuguese coastline. According to the historical records previously analyzed [1] in Lisbon municipality, the combined effects of the earthquake, tsunami, and fire caused significant damage to the city's buildings. However, the administrative limits of Lisbon municipality have been changing over time, which has been one of the limitations in the interpretation of this historical event in the city. Still, it is known that the disaster killed more than 10,000 people [1] in the municipality, which in 2010 had 54 civil parishes. Moreover, the 18th century census data show that before the earthquake Lisbon city had 109,754–157,192 residents (older than 7 years). As a result, the fatalities due to the 1755 disaster correspond to 6.4%–9.1% of the Lisbon city resident population [1]. The recovery process started immediately after the disaster. Nevertheless, only on 12 May 1758 was the Reconstruction Law of Lisbon approved [2]. It established a five-year period to conclude the reconstruction project. On the other hand, although the historical accounts are quite vast providing a general overview of the disaster in the Lisbon municipality, details remain unknown, especially in the downtown area. In addition, the 1755 event has been largely discussed among the public, stakeholders and scientific community, however, the authors did not find any published detailed analysis of the disaster in the Lisbon municipality.

Therefore, the aim of this research is to understand and reconstruct the impact of the 1755 event in downtown Lisbon, which includes earthquake, tsunami, and fire damage and the number of victims, as well as tsunami parameters (travel times, number of waves and inundation area). The study area is presented in Figure 1, corresponding to a stretch of coastline of about 1.8 km, including only 5 civil parishes (out of 54 administrative limits of Lisbon city in 2010). This area was selected due to available data and relevance to the comprehensive analysis of the tsunami. Furthermore, this research is a collaboration between academia and the Museum of Lisbon, which is quite innovative in Portugal. With this research, the authors hope to contribute to a clearer and objective understanding of this historical event in downtown Lisbon and to advance the general knowledge about this historical event that has not been properly addressed or discussed.

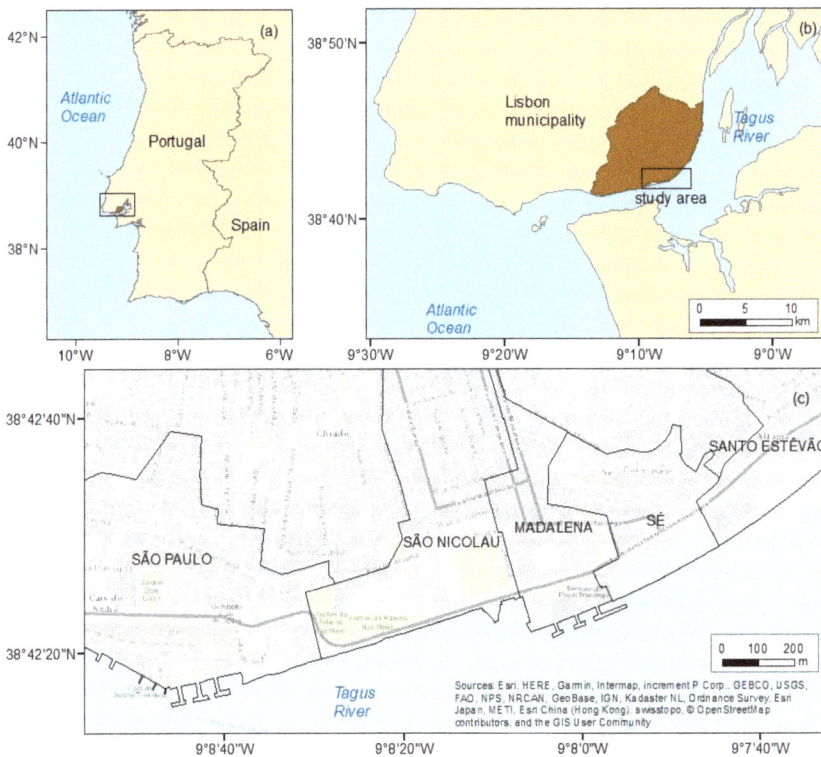

Figure 1. Geographical setting of the study area: (**a**) Location of Portugal and the Lisbon municipality; (**b**) location of the study area; (**c**) details of the study area, which in 2010 had 5 civil parishes.

2. Review of the Seismo-Tectonics Offshore of the Portuguese Mainland

The seismo-tectonics offshore of the Portuguese mainland show that the current tectonic regime at the boundary of the African and Eurasian plates is located within 35–40°N. This boundary is commonly divided into three sections [3], presented in Figure 2 ([4]): (i) the Azores Section (35°W to 24°W): the feature that dominates the Azores section is the triple point where the North American, Euroasian and African plates join. The Terceira Ridge is responsible for the active volcanism found in the Azores islands. (ii) The Central Section (24°W to 13°W): the main feature is the Gloria Fault, a rectilinear right-lateral fault. The east end of this fault is not well defined; (iii) the East section (13°W to 5°W): the tectonics of the East section (see the rectangle in Figure 2) is governed by the interaction between the Euroasian and African plates [5]. This is a region of complex bathymetry with a very low convergence rate of 4 mm yr^{-1} [6,7], trending NW–SE, consistent with the observed maximum horizontal stress direction [8–11]. The transition between the localized and diffuse plate boundary may be due to a change in the nature and age of the lithosphere [12]. The main bathymetry structures of the East section are presented in Figure 3.

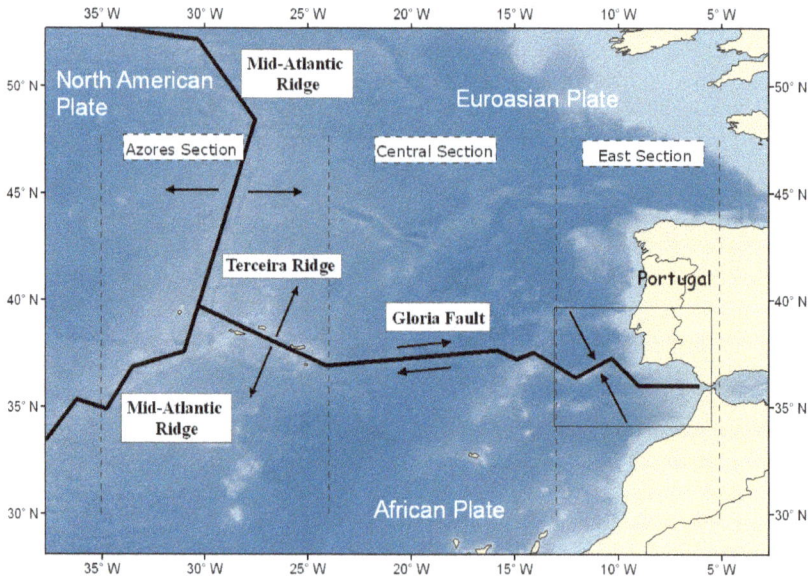

Figure 2. Plate tectonics of the Atlantic North East, and the location of the east section. The rectangle on the east section will be explained in more detail [4].

The East Section is dominated by the Gorringe Bank (Figure 3). It is orientated in a SW–NE direction, consistent with the tectonics of the region, where the stress field is dominated by a trending approximately in the NW–SE direction [5]. It has dimensions of 200 km by 80 km and reaches 25 m below the mean sea level, separating Tagus and Horseshoe Abyssal Plaines (Figure 3); it is asymmetrical, with the northern flank steeper than the southern. Gorringe Bank consists of 2 seamounts, the Gettysburg Seamount, in the Southwest and Ormonde Seamount, in the Northwest, separated by a saddle-shaped structure [13–15]. An idea of a thrust origin for the Gorringe Bank has been supported by in situ core sampling. A deep sea drilling project (DSDP) showed that the Gorringe Bank has a number of similarities to an ophiolite complex [15]. The Gettysburg Seamount appears to be essentially formed of serpentinite, whereas the basement of the Ormonde Seamount appears to consist of an oceanic section with gabbros of Berriasian age (143 Ma). It has been proposed by several authors that the Gorringe Bank is an uplifted thrust block of crustal and upper mantle rocks, resulting from the Euroasia–Africa collision (for ex., [14,15]). In fact, it has been shown that the Gorringe Bank over thrusted the Tagus Abyssal Plain for 4–5 km [16], although its seismic profile was inconclusive.

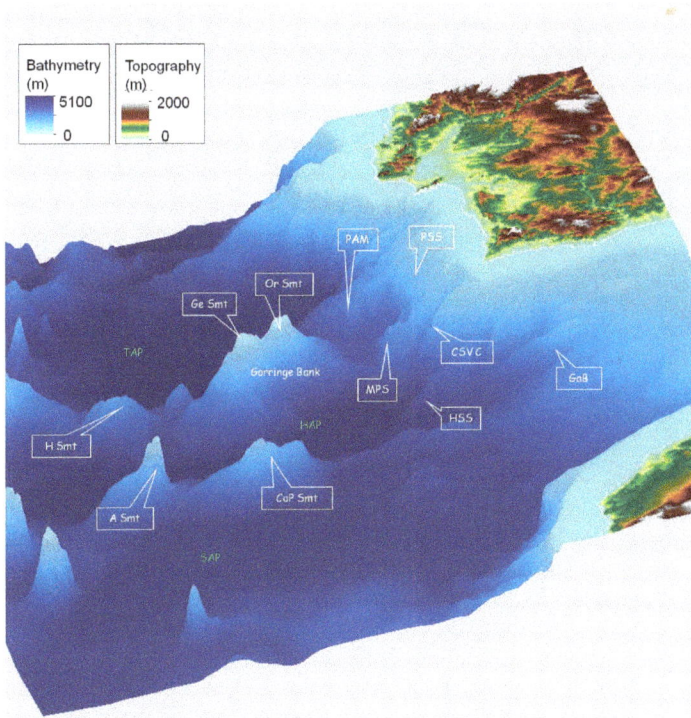

Figure 3. Portuguese continental margin represented in 3D. The most evident structure is the Gorringe Bank with 200 km in length. Abbreviations by alphabetical order are: A Smt: Ampere Seamount; CoP Smt: Coral Patch Seamount; CSVC: Cabo São Vicente Canyon; GaB: Guadalquivir Bank; Ge Smt: Gettysburg Seamount; HAP: Horseshoe Abyssal Plain; H Smt: Hirondelle Seamount; HSS: horseshoe Scarp; MPS: Marquês de Pombal Scarp; Or Smt: Ormonde Seamount; PAM: Principes de Avis Mountain; PSS: Pereira de Sousa Scarp; SAP: Seine Abyssal Plain. TAP: Tagus Abyssal Plain (adapted from [4]).

The Horseshoe Scarp (HSS) is situated to the SW of the Cabo São Vicente Canyon (CSVC) reverse fault, in the SE flank of the Horseshoe Abyssal Plain. It dips to the SE and just to the west of the fault. It is the origin of the M_s = 8.0 earthquake occurred in 28 February, 1969 (Figure 4). This was the highest instrumental earthquake event in the Iberia–Morocco–Atlantic domain [11,17].

The seismic activity at Cabo São Vicente Canyon (CSVC) could be attributed to infiltration of the sea water along the fault plane and corresponding lubrication, facilitating the movement along it [13]. Deformation is distributed over an increasingly large area that can reach a N–S width of 300 km near the continental margin of Iberia [18].

A small area between 7°W–8°W and 36°N–37°N shows significant seismic activity (Figure 4), but the bathymetry variation is not significant, compared with the previous structures. The authors did not find publications related to this area and further tectonic investigation should be carried out in this zone. Nevertheless, it is the most likely location of the source of the 1722 earthquake and tsunami (Figure 4c).

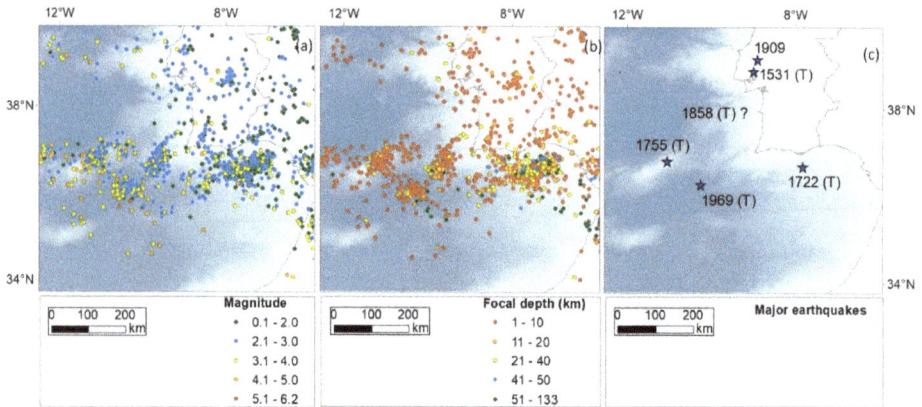

Figure 4. Seismicity of the East Section: (**a**) magnitude, (**b**) focal depth, (**c**) major earthquake. (T) indicates that the earthquake generated a tsunami. Data from United States Geological Survey (USGS), from 14 December 2000 to 14 December 2007 (adapted from [4,17]).

Another structure, but with less seismic activity, is the Pereira de Sousa Scarp (PSS), a N–S normal fault with downthrown of the west of the block. It was uplifted and rotated by ongoing compression without reactivation because it is not suitably oriented relative to the present stress field; below it is the East dipping blind thrust of Príncipe de Avis seamount. It is a possible source for the earthquake occurred on the 12th November, 1858, offshore Sines and Setúbal (Figure 4).

Guadalquivir Bank (GaB) is an E–W pop-up, 50 km long and 30 km wide. It is the source of instrumental seismicity, suggesting active uplift of an inherited paleo-relief of late Mesozoic age [8,16].

Another interesting structure is the Marquês de Pombal Scarp (MPS), which was identified as a major overthrust [18,19], oriented NNE–SSW of the continental margin over the ocean basins to the WNW, with sea bottom elevation of 1.1 km in a domain located 100 km to the SW of Portugal; this thrust could be followed in that direction for 60 km and extended at least to 30 km depth [3].

Furthermore, the instrumental seismicity of the East Section is complex and more diffuse [3,14] and the definition of the plate boundary is not clear. Generally, the earthquakes are shallow (less than 14 km), and the magnitude is less than 6.2 (Figure 4). By analyzing Figure 4, four clear clusters of earthquakes can be detected on the Gorringe Bank, on the Horseshoe Scarp (HSS), on the Cabo São Vicente Canyon (CSVC) and on a small area between 7°W–8°W and 36°N–37°N. The fact that many present-day earthquakes are situated on this section shows that the tectonic activity is still continuing, with the potential to generate large earthquakes that trigger tsunamis [20,21].

Since the 1755 earthquake and tsunami is a historical event, there are uncertainties related to the location of its source, combined with complex seismo-tectonics described above. One of the key tools to understand past tsunamis is to carry out numerical modeling, at regional scale (propagation) and local scale (inundation). A numerical model can be used to complement historical data, allowing the validation of the source model. Another advantage of this tool is to carry out simulations on coastal areas where there are no accounts. In this case, the model results provide a more comprehensive analysis of the tsunami impact. For these reasons, in this study, numerical modeling was carried out at the local scale (including inundation) at downtown Lisbon.

Moreover, several authors have used different methodologies that pointed out that the tsunami source area could be located on the Gorringe Bank (Figure 3; Figure 4): (a) Santos et al. [22] re-analyzed tsunami travel times reported by the witnesses. They conducted the wave ray analysis and determined the tsunami source area could be located on the Gorringe Bank; (b) turbidities were found nearby the Gorringe Bank by two different teams of researchers [23,24] who concluded that the records were associated with the 1755 earthquake; (c) a seismic moment assessment was conducted by [25] and the author proposed an earthquake of M_w = 8.7, with the source model located on the Gorringe Bank, and with dimensions of 200 km by 80 km; (d) seismic intensity modeling was carried out by [11], who were able to validate the observed seismic intensity; (e) numerical modeling of the 1755 tsunami was carried out at the regional scale by [22]. The authors were able to validate the tsunami initial response, which was the subsidence at Cadiz, Spain and Morocco, as reported by the witnesses. The authors also validated most of the travel times reported by the witnesses, and tsunami water level waveforms in the UK; (f) numerical modeling of the 1755 tsunami carried out at the local scale on several Portuguese coastal areas allowed researchers to calculate the tsunami inundation, reproducing tsunami travel times and local tsunami features such as the number of waves, inundation extension, and run-up [17,26–28]. As a consequence, in this study, the tsunami source area will be considered at the Gorringe Bank, with the source parameters used by [11,17,25–28]. This is a key input for the numerical modeling of the 1755 tsunami.

3. Methods

In this study several methods were used, as summarized in Figure 5. The historical data were compiled from different sources, namely historical accounts [29], and census data, both before and after the disaster [1,29]. In addition, the scale model of the city before the earthquake (Figure 6) allows a 3D perspective of Lisbon, including an indication of the buildings' heights.

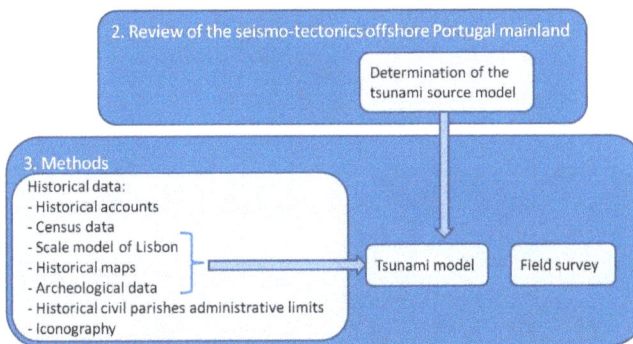

Figure 5. Schematic summary of the methods used in this study.

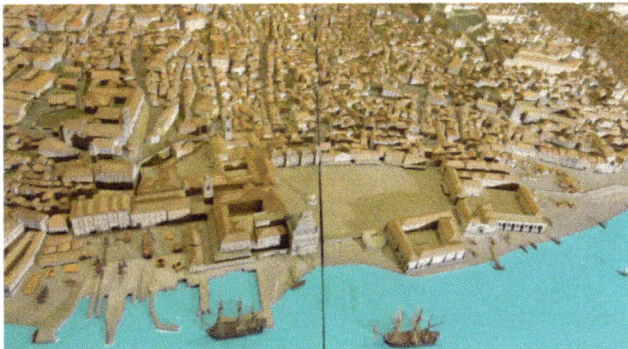

Figure 6. Part of the Lisbon city before the 1755 disaster. The scale model was built during the 1950s and is available at the Lisbon Museum to the public. Photo taken by the second author on 17 September 2018.

In order to construct the detailed digital elevation model (DEM) for the tsunami model, it was necessary to prepare the vector plot of Lisbon before the earthquake. This has been an ongoing task that started in 2005. It was based mostly on the topographic plant of the ruined city of Lisbon [30], presented in Figure 7. In this map, also the new project for the reconstruction of the city is shown [31]. The vector plot of the city before the earthquake has also been updated as new archeological evidence has been discovered [32–36]. Furthermore, archeological data was also used to identify local altimetry of the study area before the disaster, combined with modern topographic data. In addition, in order to carry out an accurate numerical model, described in more detail below, historical bathymetry maps were also compiled [37]. All the evidence shows that downtown Lisbon consisted of many open areas and squares before the earthquake. Almost all the streets were wider than 6 m, and only a few streets were narrow with the narrowest streets 3m wide. Hence, the resolution of the numerical model must reproduce these features and, accordingly, the cell size of the computational area for the inundation was 3 m.

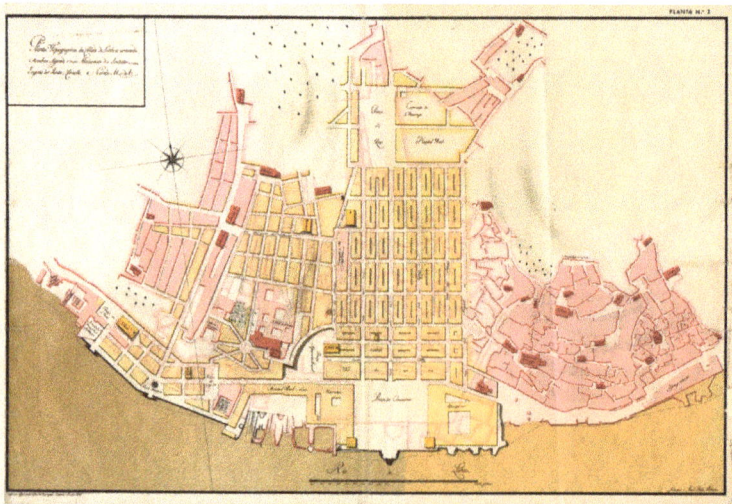

Figure 7. Reproduction of the "Topographic plant of the ruined city of Lisbon" [30].

The administrative limits of the civil parishes of Lisbon have been changing over time, and in some cases even the names have been changed. For this reason, the civil parishes' limits before the earthquake have to be compiled [38]. Iconography representing the disaster in Lisbon was also compiled. These images are available on a permanent exhibition at the Museum of Lisbon and allow a more comprehensive analysis of the disaster.

The tsunami source model considered the earthquake fault parameters proposed by [11,17,25–28], with dimensions of 200 km by 80 km, located on the Gorringe Bank. The co-seismic displacement of the seafloor is transferred to the sea surface displacement because the rupture process of an earthquake is usually much shorter than the tsunami wave period. Thus, the initial sea surface displacement of the 1755 tsunami was calculated by using the Okada formulas [39], which lead to a maximum uplift of about 6 m and a subsidence of 0.4 m (Figure 8a).

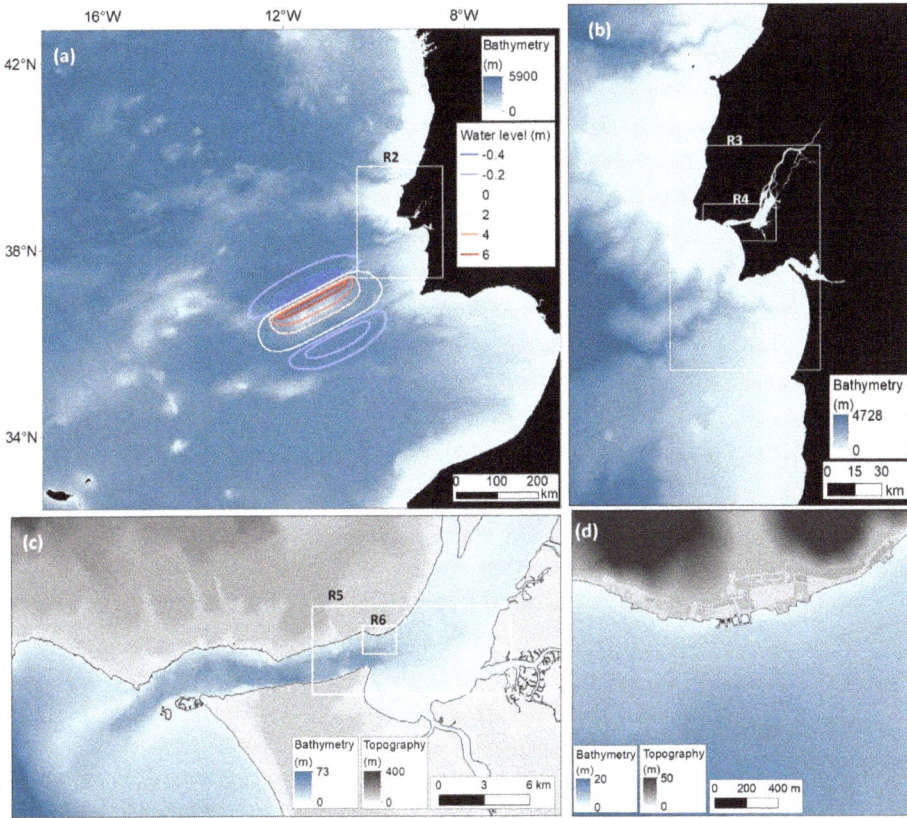

Figure 8. Conditions of the numerical model setting to downtown Lisbon: (**a**) region 1 and the initial sea surface displacement; (**b**) region 2 and the placement of regions 3 and 4; (**c**) details of region 4 and the placement of regions 5 and 6; (**d**) details of region 6.

The tsunami modeling was carried out by using the TUNAMI-N2 code of the Tohoku University which considers the non-linear shallow water equations, discretized with a staggered leap-frog finite difference scheme [40]. The governing equations, written in Cartesian coordinates, are:

$$\frac{\partial M}{\partial t} + \frac{\partial}{\partial x}\left(\frac{M^2}{D}\right) + \frac{\partial}{\partial y}\left(\frac{MN}{D}\right) + gD\frac{\partial \eta}{\partial x} + \frac{gn^2 M}{D^{\frac{7}{3}}}\sqrt{M^2 + N^2} = 0 \tag{1}$$

$$\frac{\partial N}{\partial t} + \frac{\partial}{\partial x}\left(\frac{MN}{D}\right) + \frac{\partial}{\partial y}\left(\frac{N^2}{D}\right) + gD\frac{\partial \eta}{\partial y} + \frac{gn^2 N}{D^{\frac{7}{3}}}\sqrt{M^2 + N^2} = 0 \tag{2}$$

$$\frac{\partial \eta}{\partial t} + \frac{\partial M}{\partial x} + \frac{\partial N}{\partial y} = 0 \tag{3}$$

where,

$$M = \int_{-h}^{\eta} u\,dz \tag{4}$$

$$N = \int_{-h}^{\eta} v\,dz \tag{5}$$

$$D = h + \eta \tag{6}$$

M and N are the discharge fluxes, and u and v are the velocities, in the x and y directions, respectively. D is the total water depth, η is the sea surface elevation, h is the still water depth, and g is the acceleration due to gravity. The bottom friction was considered with Manning's roughness coefficient of $n = 0.025$.

The equations were applied to a nesting of six regions, where the regions have progressively smaller areas and finer grid cell sizes and are included in the previous region, as shown in Figure 8. The first region is the largest and has a cell size of 729 m. The results of the tsunami propagation for region 1 allow a comprehensive analysis as to how the tsunami waves spread out from the tsunami source area. Then, regions 2, 3 and 4 have cell sizes of 243, 81 and 27 m, respectively. Finally, regions 5 and 6 have cell sizes of 9 and 3 m, respectively, and include details of the coastal areas and topography. During construction of each region (Figure 8), several historical bathymetry charts and topographic maps were used, as indicated above. In computational region 6, tsunami inundation allows the tsunami run-up to be calculated. In addition, in region 6, a 3 m cell size is suitable for the study area because it accurately reproduces local natural topography variations as well as structures such as buildings and streets in downtown Lisbon before the earthquake.

The results calculated in region 6 include: water level histories; snapshots of water level height; tsunami travel times, which represent the elapsed time of the first wave crest since the earthquake; inundation depth; and maximum water level.

In addition, a field survey was conducted on several occasions. The goal was to identify churches, buildings and other landmarks that still exist today on the study area. A GPS was used to georeference the collected photos. This approach was useful in the construction of the DEM of region 6 (Figure 8d). Therefore, as presented in Figure 5, the historical data, together with the numerical model results, and field survey will allow more detailed analyzes of the impact of the 1755 tsunami on the coastline of downtown Lisbon.

4. Results and Discussion

The results show the Lisbon coastline has changed significant since the earthquake, increasing the available land area, as presented in Figure 9. The coastline moved between 50 m and a maximum of about 300 m. Also presented in Figure 9 are the limits of the historical civil parishes. Although there are discrepancies between the layout of the city and the civil parishes' limits, the compilation presented in Figure 9 may be the best representation of the study area before the earthquake. On the other hand, the comparison between Figures 1 and 9 shows there is a difference between the limits of the civil parishes of São Paulo, Madalena and Sé in 2010 and before the earthquake, respectively. In addition, São Julião, São João da Praça and São Miguel civil parishes did not exist in 2010. These discrepancies make the historical interpretation even more complicated.

Therefore, it is fundamental to preserve this data for future generations. Furthermore, during the field survey, remains of the old city have been found. The small marina presented in Figure 9 is important because it shows that the vector plot of the historical map is correct and also shows that there was no significant alteration on the tide level. Thereby, the ground level remains approximately the same at about 3m above mean sea level. This was taken into consideration in the construction of the DEM for the numerical model (Figure 8d).

Figure 9. Upper plot: Lisbon city and the limits of the civil parishes, before the earthquake at the study area. These results are overlaid on the present day layout of the city. Lower plot: photo of the small marina that still exists today.

The compilation of the historical data also shows some discrepancies in the census data, before and after the disaster, as presented in Table 1. Although the records present different data, this was an attempt to identify the number of victims as well as to document the reconstruction of the city. The discrepancies are due to a variety of reasons: before the disaster, the collection of the population records was not established on credible statistical principles since the first official general census of population was carried out in 1864 (www.ine.pt). The population records were compiled by the priests during the religious ceremonies, which is a regular practice even today. Although these records are important, they do not provide a complete statistical representation of the population. Moreover, children under seven years of age were not considered by the priests. After the disaster, all the record books of São Julião civil parish were destroyed in the fire. In São João da Praça the record book of the dead was destroyed as well. Still, there are several documents that survived and in spite of the differences, discrepancies and incomplete records, it is possible to understand the impact of the disaster on each civil parish, as summarized in Table 1. The historical records show that about 1000 people died at the study area. There were fatalities in all civil parishes, except at São Miguel civil parish. In addition, the historical accounts show that the population decreased significantly because residents moved to other territories.

Table 1. Census data (before and after the earthquake) and the number of dead. Data compiled from [3]. The population data considers people older than seven years of age. The census data have several discrepancies, and different records are separated by ";".

Civil Parish	Before		After		Dead
	Buildings	**People**	**Buildings**	**People**	
São Paulo	755; 1000	4000; 7000–8000	—	4000; 1200	13; 70; 300
São Julião	1600	7016	—	1719	900
Madalena	800	3700	—	434	1; 137
Sé	896	4255	—	730	4
São João da Praça	305; 400–500	1359; 1700	10	50	2
São Miguel	870	3700	—	1850	0
Total	5226–5666	24,030–28,371	—	5983–8783	920–1343

The historical accounts are quite vast, reporting the damages of the disaster. For this reason, the damage was classified according to the overall reports and, when possible, the damages from the earthquake, tsunami and fire were analyzed separately. In general, when the buildings were not affected by each disaster, or suffered minor damage, they were classified as "no or minor damage". In several cases, the accounts report that only parts of the building were destroyed—for example, the area of the choir of a church collapsed. In these cases, these buildings were classified as "partial collapse". When the reports state everything was destroyed, these buildings were classified as "total collapse".

In spite of the vast descriptions, there are buildings that were identified but no description was found related to the 1755 disaster. As a result, this is ongoing research and, therefore, further investigation must continue; in these situations, the buildings were classified as "no information".

The earthquake effects are presented in Figure 10. The accounts show the earthquake itself did not cause significant damage in the São Julião and Madalena civil parishes, which contradicts the general knowledge, since it has been reported that the earthquake caused total destruction of the city. In addition, although the 1755 event has been largely discussed by the scientific community, stakeholders and the public, the authors did not find any publication that includes the detailed analyzes of this historical disaster in Lisbon. Therefore, this research shows that almost all the identified buildings suffered no damage or minor damage: Royal Palace (#A), Vedoria Fortress, later being only the pier, called "Cais da Pedra" (#B) and the Customs House (#C), as well as the churches (Patriarcal (#2), Madalena (#5), Padaria (#6) and Misericórdia (#7)). Other buildings that did not suffer any damage were the Palace of Corte Real (#M) and the Opera House (#N). At São Paulo and São Miguel civil parishes there was significant damage, with partial collapse of the buildings, except the Money House (#O) that also did not suffer any damage due to the earthquake. The most affected civil parishes were Sé and São João da Praça, where major damage was reported, with partial or total collapse of the buildings.

After the earthquake, the fire started and quickly spread throughout the city, being responsible for most damages to property at the entire study area (Figure 11). Only the São Miguel civil parish was not affected. Only two buildings were not affected by the fire: the Money House (#O) and São Miguel church (#10). Other buildings were affected by the fire, but were not completely destroyed (Chagas church (#11) and Palace of Conde de Coculim (#I)).

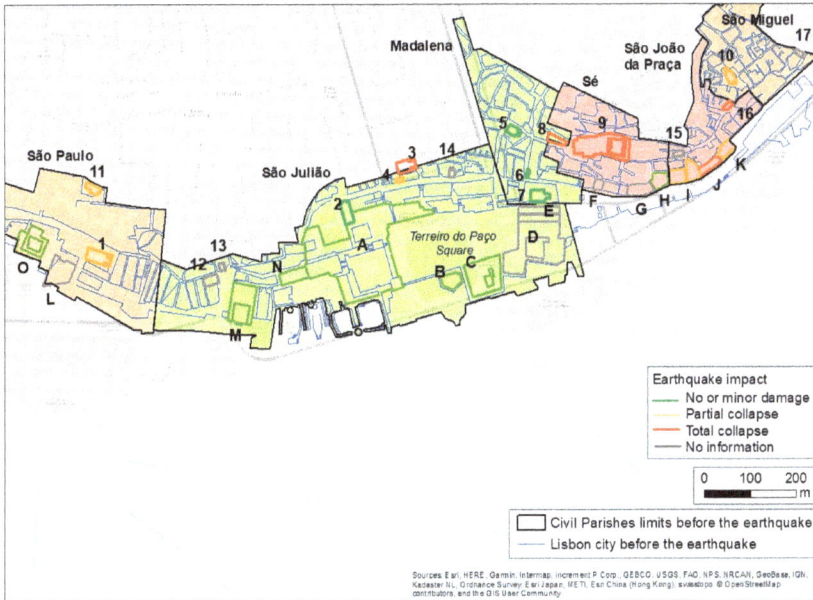

Figure 10. Damage classification caused by the earthquake on churches, other buildings and the overall damage on each civil parish at the study area. Churches: 1—São Paulo; 2—Patriarcal; 3—São Julião; 4—Ermida de Nossa. Senhora da Oliveira; 5—Madalena; 6—Ermida de São Sebastião da Padaria; 7—Misericórdia; 8—Santo António da Sé; 9—Sé; 10—São Miguel; 11—Chagas; 12—Convento dos Dominicanos Irlandeses; 13—Ermida de Nossa Senhora da Graça; 14—Nossa Senhora da Conceição; 15—São João da Praça; 16—São Pedro; 17—Ermida da Nossa Senhora dos Remédios. Other buildings: A—Paço Real da Ribeira; B—Vedoria Fortress; C—Customs House (Alfândega do Tabaco); D—Customs and Court (Alfândega e Tribunal das Sete Casas); E—Terreiro do Trigo; F—House of the Pointed Stones; G—Palace of Conde de Aveiro; H—Palace of Marquês de Távora; I—Palace of Conde de Coculim; J—Palace of Conde de Vila-Flor; K—Palace of Marquês de Angeja; L—São Paulo Fortress; M—Palace of the Corte Real; N—Opera House; O—Money House.

In addition, the tsunami was reported at São Paulo civil parish (Figure 12). The witnesses' accounts report people took refuge in the São Paulo church (#1) to escape the tsunami, but they were trapped inside the church and about 32 people died because of the fire. The church itself had already suffered partial collapse due to the earthquake (Figure 10), where an unknown number of people died. This shows that people were in panic and did not know what to do. Taking refuge inside a large church would be a common sense attitude in the 18th century. However, if people had evacuated in the direction of Chagas church (#11) they would have been safe from the tsunami (because it is located on higher ground, above 10 m) and probably the fire (because this was the limit of the burnt area). The tsunami was also reported (Figure 12) to have inundated the beaches (low ground areas) and at São Julião civil parish inundating the Terreiro do Paço square, and destroying the Vedoria Fortress, that later became a pier called Cais da Pedra (#B) and the Customs House (#C). At Terreiro do Paço more than 900 people died because they were caught by the tsunami when they tried to escape the fire. Although an open square might be a safe area to evacuate from a fire, if it is not located on high ground it could be exposed to a tsunami. Other Portuguese cities have similar urban fabric characteristics, such as an old town with narrow streets located on low ground areas. Therefore, in order to mitigate tsunamis and other urban disasters, it is fundamental to implement emergency plans, which include installation of emergency equipment, disaster awareness activities for residents and the regular practice of evacuation exercises and drills [17]. A previous study showed that in Lisbon municipality the disaster killed more than 10,000 people due to the earthquake, fire, and tsunami [1]. However, as presented in Table 1, about 1000 people died at the downtown area. These results show that the tsunami was quite relevant to the overall disaster impact on downtown Lisbon.

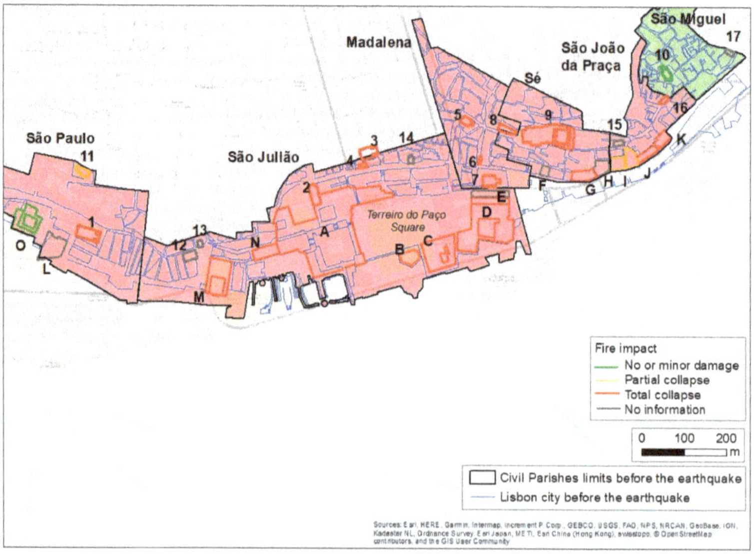

Figure 11. Damage classification caused by the fire on churches, other buildings and the overall damage on each civil parish at the study area. The list of churches and buildings is given in the caption of Figure 10.

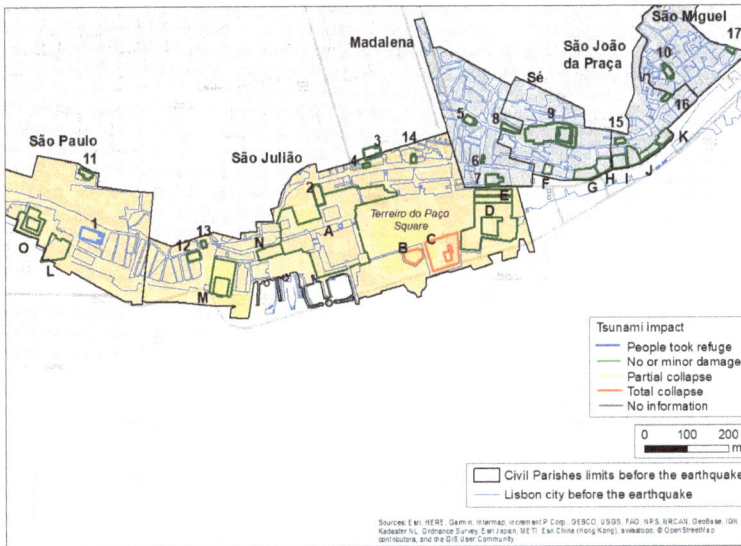

Figure 12. Damage classification caused by the tsunami on churches, other buildings and the overall damage on each civil parish at the study area. The list of churches and buildings is given in the caption of Figure 10.

The accounts are contradictory in relation to the number of waves at São Julião civil parish: some witnesses reported 3 major waves affected the Terreiro do Paço, while others observed 2 major waves. In the other civil parishes (Madalena, Sé, São João da Praça and São Miguel) there were no records about the tsunami.

The field survey shows that the House of the Pointed Stones (F) still exists today (Figure 13). It was reconstructed on the same place (Figures 10–12). Although no information was found from the historical data, onsite information indicates the third and fourth floors of the building were destroyed during the 1755 disaster. This would be classified as a "partial collapse", although no indication was provided if the damage was due to the earthquake, fire, or both.

Figure 13. House of the Pointed Stones (#F), reconstructed after the 1755 disaster.

On the other hand, the iconography allows another perspective of the disaster on the study area. Figure 14 shows the Royal Palace (#A), with view from the Terreiro do Paço. As reported by the historical accounts, the earthquake caused minor damage to the building. In addition, the drawing shows that even after the fire the building itself did not collapse. Similarly, the Opera House (#N) did not suffer significant damages due to the earthquake. However, the fire caused the total collapse of the building, and only the major stone columns remained. The São Paulo church (#1) suffered partial collapse due to the earthquake. It was then impacted by the tsunami, although no information was found about its impact on the church. Later, the building was completely destroyed by the fire. The Sé (9) was severely destroyed by both the earthquake and fire.

The tsunami model results show the first wave arrived to the study area 60 min after the earthquake (Figure 15a), inundated the low ground areas of the São Paulo civil parish. It reached up to 4.7 m high which is significant, inundating the São Paulo church (#1). This is in agreement with the witnesses reports. The first wave continued to inundate most parts of the coastline areas from 65 min onwards (Figure 15b), including the narrowest streets, the Royal Palace (A) and Terreiro do Paço square. The second wave reached the area at about 90 min but did not inundate the lower ground areas, since its height was limited to about 2 m. The third wave arrived at 130 min. In addition, some witnesses have reported 2 major waves, while others reported 3 major waves, therefore validating the numerical model results.

Figure 14. Iconography of the 1755 disaster at the study area: (**a**) copy of the drawing of the noblest part of the Royal Palace (#A) [41]; (**b**) painting of the Opera House (#N) [42]; (**c**) painting of the São Paulo church (#1) [43]; (**d**) painting of the Sé (#9) [44].

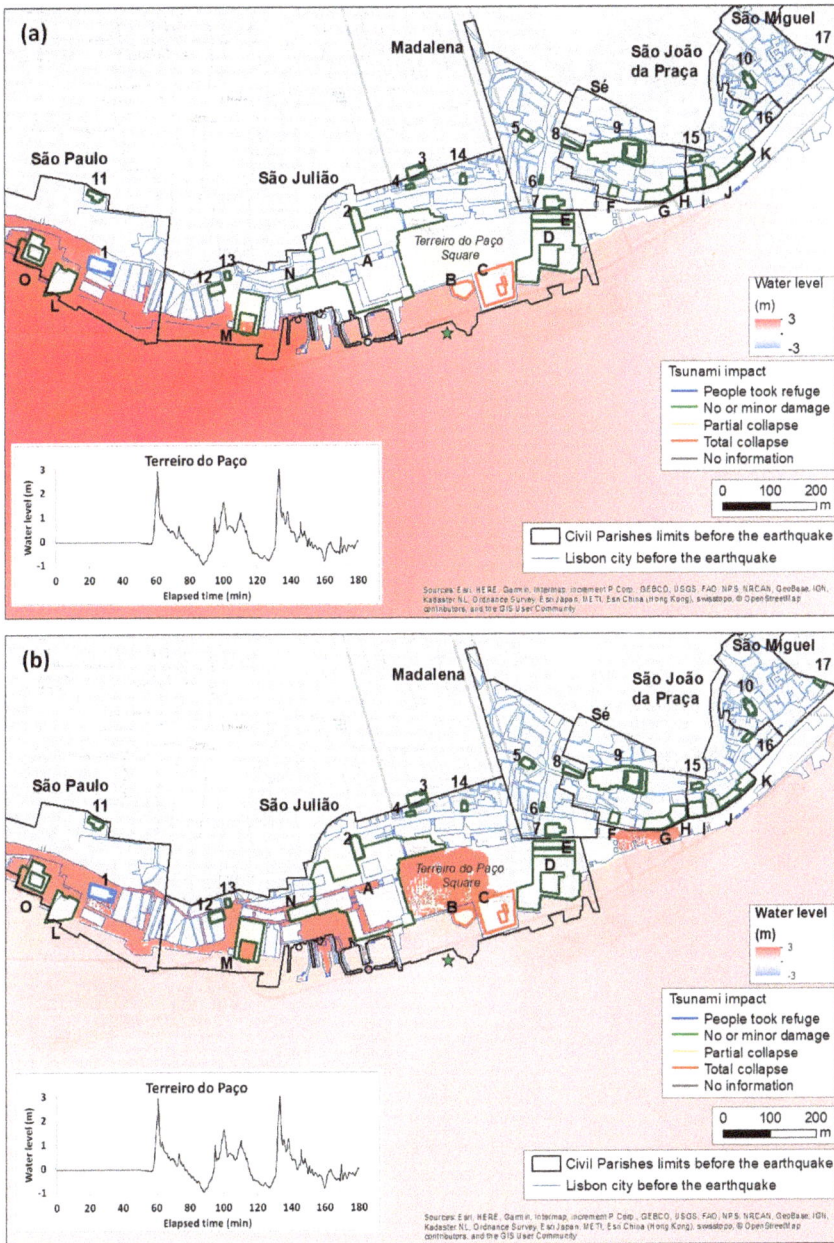

Figure 15. Numerical model results: (**a**) arrival of the first wave at 60 min; (**b**) first wave inundates most parts of the river frontside at 65 min; the embedded graphic showed the water level history in the first three hours after the earthquake at Terreiro do Paço (location of the virtual tide gauge indicated with a green star). The list of churches and buildings is given in the caption of Figure 10.

Another model output is the inundation depth, presented in Figure 16. The results show the tsunami inundated almost all the lower ground area of the downtown Lisbon coastline. The inundation depth on the São Paulo church (#1) was about 1.4 m high, at the Royal Palace (#A) it reached up to 1.0 m high, and in the Pier (Cais da Pedra) (#B) and the Customs House (Alfândega do Tabaco) (#C) it reached up to 3.0 m high. According to witnesses accounts the Customs House (#C) was destroyed by the fire and the tsunami. In the reconstruction plan of the city, it was relocated to a new location, about 650 m to the east, and still exists nowadays (Figure 16). The model results show that this area was not inundated by the tsunami. Although there are no records reporting the tsunami inundation at the São João da Praça and São Miguel civil parishes, the numerical model shows the new location for the Customs is indeed a safe place from a tsunami of this height. All these results validate the tsunami model results as well as the tsunami source model, which considered an earthquake with magnitude $M_w = 8.7$ at the Gorringe Bank. Furthermore, the new Customs building has a unique reinforcement of the facade (Figure 16) which is facing the Tagus river. This is a unique feature that was not observed on the other facades of the building, or in any other building. These results show the Portuguese stakeholders took special care in the reconstruction of the city, by providing the building with extra structural features that would ensure the safety of the building for a future tsunami.

Figure 16. Numerical model results: inundation depth. Photo of the New Customs Building (Alfândega Jardim do Tabaco). Photo taken by the first author on 23 March 2016. Dashed white lines point out the reinforcement made on the building on the river side's facade. This feature is unique in all the reconstructed buildings because it is not found on any other facade. The list of churches and buildings is given in the caption of Figure 10.

The maximum water level (Figure 17) shows the tsunami had a significant impact on the river, since it reached a height of 3.9 m at the shipyard and the marina. The maximum calculated value was 5.6 m on the west part of the city and the minimum height was 1.7 m on the east part of the city, on the river. The average value was 2.5 m. These values are sufficient to cause damage to the vessels that were on the river.

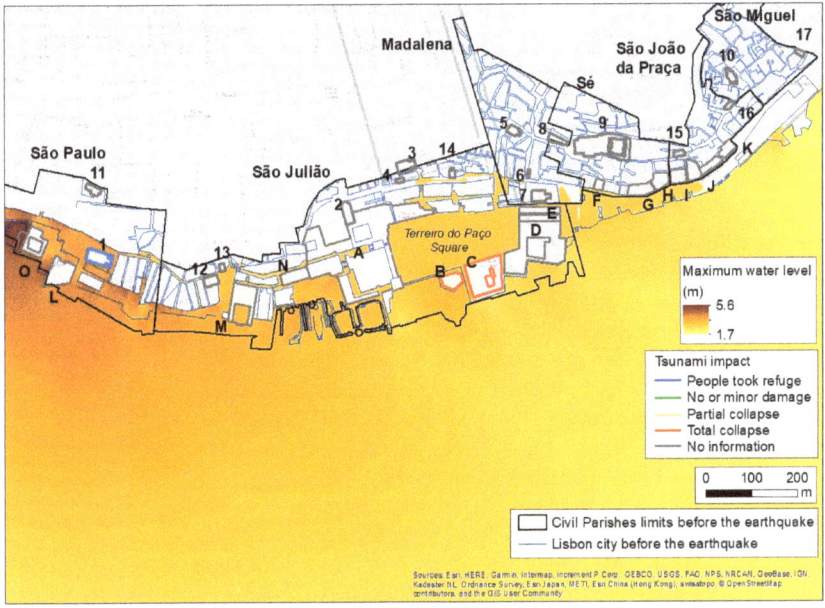

Figure 17. Numerical model results: maximum water level. The list of churches and buildings is listed in the caption of Figure 10.

The analysis of fire propagation is beyond the scope of this research. Still, from the tsunami model results it is possible to identify some hints of the fire behavior. The first wave hit São Paulo church before the fire; this means the fire reached the church later than 60–65 min after the earthquake. On the other hand, at Terreiro do Paço, the fire started first, meaning the fire arrived in the area before 65 min. Although these are rough indications, the fire may have started in the vicinity of the Terreiro do Paço, at about 1 hour after the earthquake and then propagated to the western part of the city. Still, a detailed study focusing on the fire should be conducted in the future. This is particularly important because on 28 August 1988, there was a large fire in Chiado, Lisbon. The fire caused two deaths, and injured more than 130 firefighters and local people. More than 300 people were evacuated and 18 buildings were totally destroyed. The area was renovated with modern concrete buildings and several safety improvements were implemented [45]. This shows that the city is still vulnerable to urban fires.

5. Conclusions

The aim of this research was to understand and reconstruct the impact of the 1755 event in downtown Lisbon. In order to achieve this, several methods were used, namely the compilation and interpretation of historical data, tsunami modeling, and field surveys. The combination of these methods allowed a more detailed analysis of the damage caused by the earthquake, tsunami, and fire, number of victims, as well as tsunami parameters (travel times, number of waves and inundation area).

This research showed that the coastline of downtown Lisbon has changed significantly since the earthquake, increasing the available land area by 50 m to about 300 m. In the reconstruction process of the city, the urban fabric also suffered a significant change. In addition, the administrative limits of the city's civil parishes have been changing over time. In some cases, the civil parishes have been merged or separated, which made the interpretation of the historical accounts even more difficult. Thus, the preservation of historical maps is very important in order to understand what the city looked like before the earthquake. The scale model available at the Museum of Lisbon also provided important input to reproduce the DEM of the city before the disaster.

On the other hand, census data of the studied area presented many discrepancies due to a lack of appropriate record compilation of the population. In addition, some documents were destroyed due to the fire. Still, it was possible to conclude that before the disaster there were about 5500 buildings and about 26,200 residents in downtown Lisbon; after the disaster, no records of the building were found, and there were about 6000–8800 residents. There were about 1000 deaths, mainly due to the tsunami because people evacuated to the Terreiro do Paço square, instead of moving to higher ground. Nevertheless, the number of residents decreased significantly in the study area because they moved to other territories.

Thirty two buildings were identified, among them 17 churches. In spite of the vast descriptions provided by historical data, there are buildings that were identified but no description was found relating to the 1755 disaster. For that reason, further investigation must continue.

Contradicting the general knowledge, it was possible to separate most of the buildings' damage related to the earthquake, tsunami, and fire on each historical civil parish. Moreover, the accounts show the earthquake itself did not cause significant damage in the São Julião and Madelena civil parishes, which also contradicts the general knowledge. However, at Sé and São João da Praça civil parishes, the earthquake destroyed almost all the buildings. At São Paulo and São Miguel civil parishes, the earthquake caused a partial collapse of the buildings. Moreover, the fire affected the entire study area, except São Miguel civil parish, causing the total collapse of the buildings.

The numerical model results show that the tsunami affected mostly the western and central parts of the study area and there were two major waves, which were validated by historical accounts. The model results also showed that the location of the new Customs House is in a safe area from a tsunami inundation of a similar height. These results showed that the Portuguese stakeholders took special care in the reconstruction of the city, by providing this building with extra structural features that would ensure its safety in a future tsunami. Although more than 260 years have passed since the event, there is no operational tsunami warning system in downtown Lisbon at present. As a consequence, it is fundamental to conduct tsunami awareness activities and implement evacuation strategies in order to inform residents and tourists the safest areas to evacuate in a future event.

Author Contributions: The methodology was developed by all authors; vector plant of Lisbon before the earthquake, C.L.; tsunami model, A.S.; figures and photos, A.S. and M.C.; All authors contributed with references and to the overall structure of the paper. A.S. wrote the paper, which was discussed and reviewed by all the other authors.

Funding: This research was funded by the FCT—Foundation of Science and Technology, UID/GEO/00295/2019.

Acknowledgments: The authors would like to thank the Institute of Geography of the Portuguese Army for providing topographic and land marks data. The Museum of Lisbon hereby declare that the Journal of Marine Science and Engineering is authorized to publish digital images of the historical documents listed below [30,41–44], merely as illustrations. The copyrights of the referred documents belong exclusively to the Museum of Lisbon.

Conflicts of Interest: The authors declare no conflict of interest.

References

1. Santos, A.; Koshimura, S. The historical review of the 1755 Lisbon Tsunami. *J. Geod. Geomat. Eng.* **2015**, *1*, 38–52.

2. França, J.A. *A Reconstrução de Lisboa e a Arquitectura Pombalina*; Instituto de Cultura e Língua Portuguesa Ministério da Educação, Biblioteca Breve: Lisboa, Porugal, 1989; Volume 12, p. 30. (In Portuguese)

3. Buforn, E.; Udias, A.; Colombas, M.A. Seismicity, source mechanisms and tectonics of the Azores-Gibraltar plate boundary. *Tectonophys.* **1988**, *152*, 89–118. [CrossRef]

4. Santos, A. Tsunami hazard assessment in Portugal by the worst case scenario: The November 1st, 1755 Lisbon Tsunami. Ph.D. Thesis, Tohoku University, Sendai, Japan, 2008; p. 210.

5. Moreira, V.S. Seismotectonics of Portugal and its adjacent area in the Atlantic. *Tectonophysics* **1985**, *11*, 85–96. [CrossRef]

6. DeMets, C.; Gordon, R.G.; Argus, D.F.; Stein, S. Current plate motions. *Geophys. J. Int.* **1990**, *101*, 425–478. [CrossRef]

7. McClusky, S.; Reilinger, R.; Mahmoud, S.; Ben Sari, D.; Tealeb, A. GPS constraints on Africa (Nubia) and Arabia plate motions. *Geophys. J. Int.* **2003**, *155*, 126–138. [CrossRef]
8. Ribeiro, A.; Mendes-Victor, L.; Cabral, J.; Matias, L.; Terrinha, P. The 1755 Lisbon earthquake and the beginning of closure of the Atlantic. *Eur. Rev.* **2006**, *14*, 193–205. [CrossRef]
9. Borges, J.F.; Fitas, A.J.S.; Bezzeghoud, M.; Teves-Costa, P. Seismotectonics of Portugal and its adjacent Atlantic area. *Tectonophysics* **2001**, *337*, 373–387. [CrossRef]
10. Carrilho, J.; Teves-Costa, P.; Morais, I.; Pagarete, J.; Dias, R. GEOALGAR project: First results on seismicity and fault-plane solutions. *Pure Appl. Geophys.* **2004**, *161*, 589–606. [CrossRef]
11. Grandin, R.; Borges, J.F.; Bezzeghoud, M.; Caldeira, B.; Carrilho, F. Simulations of strong ground motion in SW Iberia for the 1969 February 28 (Ms = 8.0) and the 1755 November 1 (M ~ 8.5) earthquakes—II. Strong ground motion simulations. *Geophys. J. Int.* **2007**, *171*, 807–822. [CrossRef]
12. Jimenez-Munt, I.; Fernandez, M.; Torne, M.; Bird, P. The transition from linear to diffuse plate boundary in the Azores-Gibraltar region: Results from a thin-sheet model. *Earth Planet Sci. Lett.* **2001**, *192*, 175–189. [CrossRef]
13. Moreira, V.S. Seismicity of the Portuguese continental margin. In *Earthquakes at North-Atlantic Passive Margins: Neotectonics and Postglacial Rebound*; Gregersen, S., Basham, P.W., Eds.; NATO ASI Series, Series C; Springer: New York, NY, USA, 1989; Volume 266, pp. 533–545.
14. Pinheiro, L.M.; Wilson, R.C.L.; dos Reis, R.P.; Whitmarsh, R.B.; Ribeiro, A. The Western Iberia Margin: A Geophysical and Geological Overview. In Proceedings of the Ocean Drilling Program, Scientific Results, Lisbon, Portugal, 10 March–29 May 1993; Whitmarsh, R.B., Sawyer, D.S., Klaus, A., Masson, D.G., Eds.; National Science Foundation: Alexandria, VA, USA, 1996; Volume 149, p. 23.
15. Hayward, N.; Watts, A.B.; Westbrook, G.K.; Collier, J.S. A seismic reflection and GLORIA study of compressional deformation in the Gorringe Bank region, eastern North Atlantic. *Geophys. J. Int.* **1999**, *138*, 831–850. [CrossRef]
16. Zitellini, N.; Rovere, M.; Terrinha, P.; Chierici, F.; Matias, L.; Bigsets Team. Neogene through quaternary tectonic reactivation of SW Iberian passive margin. *Pure Appl. Geophys.* **2004**, *161*, 565–587. [CrossRef]
17. Santos, A.; Fonseca, N.; Queirós, M.; Zezere, J.L.; Bucho, J.L. Implementation of tsunami evacuation maps at Setubal Municipality, Portugal. *Geosciences* **2017**, *7*, 116. [CrossRef]
18. Zitellini, N.; Mendes, L.A.; Bigsets Team. Source of 1755 Lisbon earthquake and tsunami investigated. *EOS Trans. Ame. Geop. Un.* **2001**, *82*, 285, 290–291.
19. Grimison, N.L.; Chen, W.P. The Azores-Gibraltar plate boundary: Focal mechanisms, depths of earthquakes, and their tectonic implications. *J. Geop. Res.* **1986**, *91*, 2029–2047. [CrossRef]
20. Gutscher, M.A. What caused the Great Lisbon earthquake? *Science* **2004**, *305*, 1247–1248. [CrossRef] [PubMed]
21. Barkan, R.; Sten, U.; Brink, T.; Lin, J. Far field tsunami simulations of the 1755 Lisbon earthquake: Implications for tsunami hazard to the U.S. East Coast and the Caribbean. *Mar. Geol.* **2009**, *264*, 109–122. [CrossRef]
22. Santos, A.; Koshimura, S.; Imamura, F. The 1755 Lisbon Tsunami: Tsunami source determination and its validation. *J. Disaster Res.* **2009**, *4*, 41–52. [CrossRef]
23. Lebreiro, S.M.; McCave, I.N.; Weaver, P.P.E. Late quaternary turbidite emplacement on the Horseshoe Abyssal Plain (Iberian Margin). *J. Sediment. Res.* **1997**, *67*, 856–870.
24. Thomson, J.; Weaver, P.P.E. An AMS radiocarbon method to determine the emplacement time of recent deep-sea turbidites. *Sediment. Geol.* **1994**, *89*, 1–7. [CrossRef]
25. Johnston, A. Seismic moment assessment of earthquakes in stable continental regions–III. New Madrid 1811–1812, Charleston 1886 and Lisbon 1755. *Geophys. J. Int.* **1996**, *126*, 314–444. [CrossRef]
26. Santos, A.; Pereira, S.; Fonseca, N.; Paixão, R.; Andrade, F. Tsunami risk assessment in the Peniche and Lourinha municipalities. In *The 1755 Lisbon Earthquake: What Have We Learned*; Imprensa da Universidade de Coimbra: Coimbra, Portugal, 2015; pp. 251–276. (In Portuguese)
27. Santos, A.; Koshimura, S. The 1755 Lisbon tsunami at Vila do Bispo municipality, Portugal. *J. Disaster Res.* **2009**, *10*, 1067–1080. [CrossRef]
28. Santos, A.; Queirós, M.; Rodriguez, J. Tsunami risk assessment at Albufeira downtown, Portugal. In *Proceedings of the ICUR 2016, Lisbon, Portugal, 30 June–2 July 2016*; Costa, P., Quin, D., Garcia, R., Eds.; Centro Europeu de Riscos Urbanos (EUR-OPA): Lisbon, Portugal, 2016; pp. 449–454. ISBN 978-989-95094-1-2.
29. Sousa, L. *The Earthquake of the November 1, 1755 and a Demographic Study*; Portuguese Geological Services: Lisbon, Portugal, 1928; Volume 3, p. 1013. (In Portuguese)

30. Ribeiro, J.P. *Reprodução da Planta Topographica da Cidade de Lisboa arruinada, Tambem Segundo o novo Alinhamento dos Architétos, Eugénio dos Santos, e Carvalho, e Carlos Mardel Projeto escolhido para a reconstrução de Lisboa após o Terramoto de 1755 e datado de 12 de junho de 1758. Gravura, Litografia*; Museu de Lisboa, EGEAC: Lisbon, Portugal, 1949. (In Portuguese)

31. Anonymous. *Planta da ribeira da Cidade de Lisboa thé Santos*; Desenho, Tinta da China e Aguarela, Primeira metade do sec. XVIII; Museu de Lisboa, EGEAC: Lisbon, Portugal. (In Portuguese)

32. Santos, M.J. The Largo Vitorino Damásio (Santos o Velho), Lisbon: Contribution to the history of the riverside area of Lisbon. *Port. J. Archeol.* **2006**, *9*, 369–399.

33. Blot, M.L.P.; Henriques, R. Urban archeology and archeology of the aquatic environment. The port problem as a 'bridge' between two research territories. In *História, Teoria e Método da Arqueologia. Proceedings of the IV Congresso de Arqueologia Peninsular*; Faculdade de Ciências Humanas e Sociais da Universidade do Algarve: Faro, Portugal, 14–19 September 2011; Volume 14, pp. 127–140. (In Portuguese)

34. Macedo, M.L.; Silva, I.M.; Lopes, G.C.; Bettencourt, J. The maritime dimension of the Boqueirão do Duro (Santos, Lisboa) in the 18th and 19th centuries: First archaeological results. In *Arqueologia em Portugal*; Morais Arnaud, J., Martins, A., Eds.; Estado da Questão; AAP: Lisboa, Portugal, 2017; pp. 1915–1924. (In Portuguese)

35. Macedo, M.L.; Sarrazola, A. *Parking lot of Praça, D. Luís, I*; ERA—Arqueologia, S.A.: Lisbon, Portugal, 2012; pp. 34–62. (In Portuguese)

36. Bettencourt, J.; Carvalho, P.; Fonseca, C.; Coelho, I.P.; Lopes, G.; Silva, T. *Ships of the modern period in Lisbon: Balance and research perspectives. In I Encontro de Arqueologia de Lisboa: Uma Cidade em Escavação*; Nozes, C., Cameira, I., Banha da Silva, R., Eds.; Livro de Resumos: Lisbon, Portugal, 2017; Caessa, A.; pp. 478–495. (In Portuguese)

37. Portuguese Navy. *Entrance of the river Tagus with the harbour of Lisbon*; Portuguese Navy: Lisbon, Portugal, 1879.

38. Lisbon City Hall. *Streets and Limits of the Civil Parishes of Lisbon*; Torre do Tombo: Lisbon, Portugal, 1770; Volume 153. (In Portuguese)

39. Okada, Y. Surface deformation due to shear and tensile faults in a half space. *Bull. Seism. Soc. Am.* **1985**, *75*, 1135–1154.

40. Imamura, F. Review of tsunami simulation with a finite difference method. In *Long-Wave Runup Models*; World Scientific: Singapore, 1995; pp. 25–42. (In Portuguese)

41. Da Cunha, L. *Copia do Desenho Parte Mais Nobre do Palacio do Rey de Portugal, segunda metade do século XVIII*; Tinta da China e Aguada, D.T., João, V.F., Eds.; Museu de Lisboa, EGEAC: Lisbon, Portugal, 1922. (In Portuguese)

42. Le Bas, J.P.; Pedegache, M.T. *Casa da Opera, Gravura Aguarelada*; Museu de Lisboa, EGEAC: Lisbon, Portugal, 1757. (In Portuguese)

43. Le Bas, J.P.; Pedegache, M.T. *Igreja de São Paulo, Gravura Aguarelada*; Museu de Lisboa, EGEAC: Lisbon, Portugal, 1757. (In Portuguese)

44. Le Bas, J.P.; Pedegache, M.T. *Basilica de Santa Maria, Gravura Aguarelada*; Museu de Lisboa, EGEAC: Lisbon, Portugal, 1757. (In Portuguese)

45. Santos, A.; Queirós, M.; Carvalho, L. Fire and seismic risk perception at Lisbon University—Faculty of Letters. *Territorium* **2017**, *24*, 15–27. [CrossRef]

© 2019 by the authors. Licensee MDPI, Basel, Switzerland. This article is an open access article distributed under the terms and conditions of the Creative Commons Attribution (CC BY) license (http://creativecommons.org/licenses/by/4.0/).

Journal of
Marine Science and Engineering

MDPI

Article

Impulse Wave Runup on Steep to Vertical Slopes

Frederic M. Evers * and Robert M. Boes

Laboratory of Hydraulics, Hydrology and Glaciology (VAW), ETH Zurich, CH-8093 Zürich, Switzerland;
boes@vaw.baug.ethz.ch
* Correspondence: evers@vaw.baug.ethz.ch; Tel.: +41-44-633-0877

Received: 9 November 2018; Accepted: 18 December 2018; Published: 7 January 2019

Abstract: Impulse waves are generated by landslides or avalanches impacting oceans, lakes or reservoirs, for example. Non-breaking impulse wave runup on slope angles ranging from 10° to 90° (V/H: 1/5.7 to 1/0) is investigated. The prediction of runup heights induced by these waves is an important parameter for hazard assessment and mitigation. An experimental dataset containing 359 runup heights by impulse and solitary waves is compiled from several published sources. Existing equations, both empirical and analytical, are then applied to this dataset to assess their prediction quality on an extended parameter range. Based on this analysis, a new prediction equation is proposed. The main findings are: (1) solitary waves are a suitable proxy for modelling impulse wave runup; (2) commonly applied equations from the literature may underestimate the runup height of small wave amplitudes; (3) the proposed semi-empirical equations predict the overall dataset within ±20% scatter for relative wave crest amplitudes ε, i.e., the wave crest amplitude normalised with the stillwater depth, between 0.007 and 0.69.

Keywords: impulse wave; solitary wave; landslide tsunami; wave runup; runup prediction

---

1. Introduction

Impulse waves are generated by very rapid gravity-driven mass movements including landslides and avalanches impacting a body of water (Heller et al. [1]). The slide energy is transferred to the water column and a wave train is generated, which propagates away from the impact location. Especially water bodies with steep shorelines, e.g., fjords, mountain lakes or reservoirs, are prone to this tsunami-like hazard. Roberts et al. [2] compiled a global catalogue with 254 landslide-generated impulse wave events. In the past, extreme absolute impulse wave runup heights were observed in Lituya Bay, USA, in 1958 with 524 m (Miller [3]), Chehalis Lake, Canada, in 2007 with 38 m (Roberts et al. [4]), and Taan Fjord, USA, in 2015 with 193 m (Higman et al. [5]).

For hazard mitigation, the runup height R is of primary interest (Figure 1). While the impulse wave events given above are extreme cases, comparably small runup heights at densely populated lake shores may already cause substantial damage (Fuchs and Boes [6]). Particularly in reservoirs, where there is a freeboard of just a few meters between the stillwater level and the dam crest, the prediction of runup by small impulse wave amplitudes needs to be as accurate as possible to prevent overtopping. Müller [7] conducted experiments specifically designed to study the runup of impulse waves and derived an empirical prediction equation for the runup height induced by the first or leading wave, respectively, of the impulse wave train. However, also equations derived from experiments with solitary waves are commonly applied to predict runup heights by impulse waves (e.g., Bregoli et al. [8], McFall and Fritz [9]). While a solitary wave features a single wave crest, impulse waves are characterized by an outgoing wave train with multiple wave crests and troughs (Figure 1).

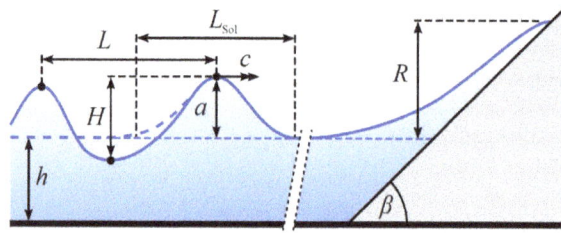

Figure 1. Definition plot for impulse (solid line) and solitary wave (dashed line) runup on an inclined slope.

This study focuses on non-breaking impulse wave runup on steep slopes, i.e., slope angles $\beta \geq 10°$ (Figure 1). First, runup equations from the literature are discussed. Since there is a multitude of studies on solitary wave runup on gentle slopes (Pujara et al. [10], Hafsteinsson et al. [11]), only equations derived from experiments with $\beta \geq 10°$ or those commonly applied for this parameter range are included. A dataset consisting of measured runup heights from both impulse wave and solitary wave (index Sol) runup experiments with β between 10° and 90° (vertical wall) is then compiled from published experimental data. Subsequently, the runup equations are applied to these data for runup prediction. Based on this comparison, a semi-empirical runup equation is proposed. The discussion includes the limitations of this equation and assesses the significance of scale effects.

2. Runup Equations

The governing parameters included in the prediction equations for the runup height described below are: wave crest amplitude a, wave height H, wave length L, stillwater depth h, and runup slope angle β (Figure 1). While H is a combined parameter of the first wave crest and trough amplitudes for impulse waves, $H = a$ for solitary waves. The relative wave crest amplitude is defined as $\varepsilon = a / h$. While the length of a solitary wave is infinite (Dean and Dalrymple [12]), it may be approximated with L_{Sol} as the effective wave length (Lo et al. [13]). For empirically derived prediction equations included below, the corresponding datasets are described in the following section.

Müller [7] approximated the runup height R of impulse waves based on wave channel experiments with

$$\frac{R}{h} = 1.25 \left(\frac{H}{h}\right)^{1.25} \left(\frac{H}{L}\right)^{-0.15} \left(\frac{90°}{\beta}\right)^{0.2}.$$

(1)

Equation (1) contains the two governing wave parameters, H and L. The last term including β equals to 1 for 90° and increases for decreasing β. To predict the runup height of a solitary wave, its effective wave length may be approximated with (Lo et al. [13])

$$L_{Sol} = \frac{2\pi h}{\sqrt{0.75\varepsilon}}.$$

(2)

Hall and Watts [14] approximated the runup height R of solitary waves also based on wave channel experiments with

$$\frac{R}{h} = 3.05 \left(\frac{\beta}{180°}\pi\right)^{-0.13} \varepsilon^{1.15\left(\frac{\beta}{180°}\pi\right)^{0.02}}$$

(3)

In the original publication, the runup slope angle β was expressed in radians, whereas its unit is in (°) in Equation (3). Since this equation was derived for solitary wave runup, ε is included as the single governing wave parameter. Hall and Watts [14] state the application range for Equation (3) with β between 12° and 45°. A different equation is given for β between 5° and 12°. However, the latter will not be considered in the further analysis, while Equation (3) will be applied to runup angles between 10° and 45° for simplification.

As the third empirical runup equation, Fuchs and Hager [15] approximated the runup height R of solitary waves in wave channel experiments with

$$\frac{R}{h} = 3(\tan \beta)^{-0.05} \varepsilon \tag{4}$$

The governing wave parameter is ε. The runup slope β is included in a tangent function. As the tangent of $\beta = 90°$ is not defined, wave runup at a vertical wall may not be predicted with Equation (4).

Synolakis [16] developed an approximate solution of the nonlinear wave theory and applied it to derive an equation describing the maximum runup height R of a solitary wave with

$$\frac{R}{h} = 2.831(\cot \beta)^{0.5}\varepsilon^{1.25}, \tag{5}$$

referred to as *the runup law*. Equation (5) is formally correct for $\varepsilon^{0.5} \gg 0.288 \tan \beta$ (Synolakis [16]). The theoretical approach is compared to own experimental data for a gentle slope with $\beta \approx 2.9°$ ($\tan\beta = 1/19.85$) as well as to selected experiments with non-breaking wave runup data by Hall and Watts [14] with β between 5° and 45°. Synolakis [16] finds a satisfactory agreement except for $\beta = 45°$. Similar to Equation (4), the cotangent of $\beta = 90°$ equals to zero and therefore wave runup at a vertical wall may not be predicted with Equation (5).

Except for Equation (1), none of the other equations considered runup at a vertical wall, i.e., $\beta = 90°$. Su and Mirie [17] studied the collision of two solitary waves and conducted a perturbation analysis of this phenomenon to the third order. If viscosity and surface tension effects are neglected, the case of two colliding solitary waves with the same amplitude is equal to the runup of a single wave at a vertical wall (Cooker et al. [18]). The maximum runup height R for $\beta = 90°$ is stated by Su and Mirie [17] with

$$\frac{R}{h} = 2\varepsilon + \frac{1}{2}\varepsilon^2 + \frac{3}{4}\varepsilon^3. \tag{6}$$

For very small relative wave amplitudes, the relative runup height R/h converges to 2ε. Maxworthy [19] presents experimental results for both wave-wave interaction and wave-wall interaction, which indicate that the maximum superposed wave amplitude or runup height, respectively, is approximately 10% higher in the former than in the latter case. However, these experiments were conducted at stillwater depths between 4.5 and 6.7 cm and viscosity as well as surface tension effects might have had a more significant influence compared to larger scales.

3. Datasets

Five experimental datasets were included in the analysis presented herein: Müller [7] ($n = 166$), Hall and Watts [14] ($n = 138$), Fuchs [20] ($n = 19$), Street and Camfield [21] ($n = 22$), and Maxworthy [19] ($n = 14$) with a total of $n = 359$ experiments. Their respective key parameter ranges are summarized in Table 1. In Müller's [7] experiments, wave trains with multiple wave crests and troughs were generated to reproduce landslide generated impulse wave characteristics, while the other experimental series document solitary waves, i.e., $a = H$ (Figure 1). Therefore, the waves generated in the experiments by Müller [7] will be referred to in the following as *impulse waves*, while the remaining are *solitary waves*. The generated relative wave amplitudes ε range from 0.007 to 0.69 and runup slope angles β from 10° to 90° (vertical wall). While empirical prediction equations were directly derived from the datasets by Müller [7], Hall and Watts [14], and Fuchs [15] as described in the previous section, the data by Street and Camfield [20] and Maxworthy [19] for $\beta = 90°$ is included to assess the analytically derived Equation (6) by Su and Mirie [17].

Table 1. Parameter ranges of datasets included in the analysis.

Dataset	h (m)	β (°)	ε (-)	R/h (-)	a/H (-)
Müller [7]	0.2–0.6	18.4–90	0.007–0.495	0.014–1.143	0.57–1.04
Hall and Watts [14]	0.15–0.69	10–45	0.05–0.56	0.09–1.82	1 [1]
Fuchs [20]	0.2	11.3–33.7	0.10–0.69	0.33–2.30	1 [1]
Street and Camfield [21]	0.152–0.305	90	0.10–0.65	0.18–1.75	1 [1]
Maxworthy [19]	0.045–0.067	90	0.12–0.67	0.28–1.55	1 [1]

[1] Solitary wave.

Müller [7] conducted experiments in a wave channel with a length of 19.25 m, a width of 1 m, and a depth of 1.2 m. The impulse waves were generated by a rectangular box falling vertically onto the water surface at one end of the channel. The box mass ranged from 118 to 422 kg. By adjusting box mass, drop height, and water depth, tests with differing wave characteristics were achieved. Parallel wire wave gauges were applied for measuring the water surface displacement. The installed slope angles β were 18.4°, 45° and 90°. For tracking the maximum runup height R of the first wave at the vertical wall ($\beta = 90°$), also wire wave gauges were applied. For the two milder slopes, R was optically recorded. The accuracy of the wave gauges is given with ± 0.1 mm and for the optical method with ± 1 to 2 mm. Repeatability tests yielded deviation $< 1\%$ for runup heights $R > 40$ mm and a larger scatter of 10% for $R < 40$ mm. Figure 2 shows the wave crest celerities c (see Figure 1) of the first outgoing wave within the impulse wave train as a function of ε. Compared to the celerity of a solitary wave defined by Russell [22] as:

$$\frac{c_{Sol}}{\sqrt{gh}} = \sqrt{1+\varepsilon}, \tag{7}$$

the measured impulse wave celerities mainly scatter between 95% and 103% of the solitary wave celerity. This agrees with the findings by McFall and Fritz [23] and Evers et al. [24] for spatial impulse wave propagation in wave basins. The relative wave length L/h of Müller's [7] experiments range between 9 and 56. According to Dean and Dalrymple [12], the generated waves cover the transition zone from intermediate-water ($2 < L/h \leq 20$) to shallow-water ($L/h > 20$). The measured wave celerities confirm this classification. In line with the additional text information in Müller [7], experiments no. 474, 589, 601, and 602 were not included in the analysis. The experiments featuring roughness elements at $\beta = 18.4°$ (no. 562–588) were also not considered. All other experiments from Müller's [7] appendix providing sufficient information on the impulse wave characteristics as well as the runup height were included in the analysis. The number of experiments included from Müller [7] is $n = 166$ (63 at $\beta = 18.4°$, 17 at $\beta = 45°$, 86 at $\beta = 90°$).

The experiments by Hall and Watts [14] were conducted in a wave channel with a length of 25.9 m, a width of 4.3 m, and a depth of 1.2 m. The solitary waves were generated with a "pusher type" wave generator featuring a vertical pusher face mounted to a trolley, which was mechanically linked to a gravitationally accelerated drop weight. Both, wave height and runup height were optically measured. Besides slope angles $\beta = 10°$, 15°, 25°, and 45°, also $\beta = 5°$ was installed in the channel. However, the latter was not included into this analysis. The measured wave celerities c_{Sol} are within 88% to 94% of Equation (7) (Figure 2). Two experiments were not included into the analysis for this study as they were quite outside the trend of neighboring data points and therefore classified as obvious outliers ($R/h = 0.535$ and 0.679, both at $\beta = 10°$). The number of experiments included from Hall and Watts [14] is $n = 138$ (38 at $\beta = 10°$, 37 at $\beta = 15°$, 31 at $\beta = 25°$, 32 at $\beta = 45°$).

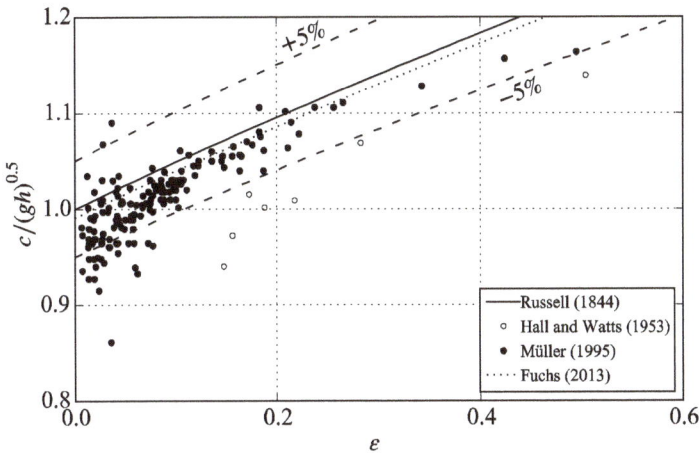

Figure 2. Relative wave crest celerity $c/(gh)^{0.5}$ versus relative wave amplitude ε from data sources from Müller [7], Hall and Watts [14], and Fuchs [20] and (—) solitary wave celerity (Equation (7)).

Fuchs [20] conducted experiments with a pneumatic piston-type wave generator in an 11 m long, 0.5 m wide, and 1 m deep channel. The inclinations of the runup slopes were set to tan β = 1/5, 1/2.5, and 1/1.5; i.e., $\beta \approx 11.3°$, 21.8°, and 33.7°. The solitary wave profiles were tracked with ultrasonic distance sensors and runup heights were measured optically. The measured wave celerities average out at 99.1% of Equation (7) (Figure 2). The number of experiments included from Fuchs [20] is $n = 19$ (7 at $\beta = 11.3°$, 6 at $\beta = 22.8°$, 6 at $\beta = 33.6°$).

Street and Camfield [21] studied solitary wave reflection at a vertical wall, i.e., $\beta = 90°$, in a 17 m long subsection of a channel with a length of 35 m and a width of 0.91 m. The piston-type wave generator was controlled with a hydraulic-servo-electronic system. Wave amplitudes were measured with capacitance wave gauges. Near shore deformation details were optically recorded. The runup data by Street and Camfield [21] was extracted with WebPlotDigitizer [25] from the original publication. The number of experiments included from Street and Camfield [21] is $n = 22$.

Maxworthy [19] studied solitary wave reflection at a vertical wall as well as head-on collision between two solitary waves in a 5 m long, 0.2 m wide, and 0.3 m deep channel. Only runup data at the wall was considered for this study. The waves were generated manually by pulling a plate through the channel. The wave characteristics and the runup height were measured optically. The runup data by Maxworthy [19] was extracted with WebPlotDigitizer [25] from the original publication. The number of experiments included from Maxworthy [19] is $n = 14$.

In total, 359 experiments were included into the analysis. The available information on wave celerities and lengths indicates that the waves contained in the dataset may be classified into the transition zone from intermediate-water to shallow-water or long waves, respectively. Figure 3a shows the runup height R over the wave amplitude a versus the slope parameter S_0 introduced by Grilli et al. [26] with

$$S_o = 1.521 \frac{\tan \beta}{\sqrt{\varepsilon}}, \tag{8}$$

for the overall dataset except for the experiments with $\beta = 90°$. S_o allows for assessing whether a solitary wave is breaking or non-breaking during runup. Grilli et al. [26] stated $S_o > 0.37$ as the criterion for non-breaking solitary wave runup. As β approaches 90°, S_o tends to infinity. Therefore, experiments with $\beta = 90°$ are not included in Figure 3. All other experiments satisfy $S_o \geq 0.37$. The experiment by Fuchs [20] with $S_o = 0.37$ (Figure 3), $\beta \approx 11.3°$, and $\varepsilon = 0.69$ is shown in Figure 4 and features no distinct wave breaking characteristics. It is therefore assumed that the overall dataset consists of non-breaking wave runup. The limiting criterion $\varepsilon^{0.5} \gg 0.288 \tan\beta$ for Equation (5) stated by Synolakis [16] may be reformulated as $S_o \ll 5.28$ (Pujara et al. [10]) and is also included in Figure 3. Several experiments feature S_o values close to and larger than 5.28.

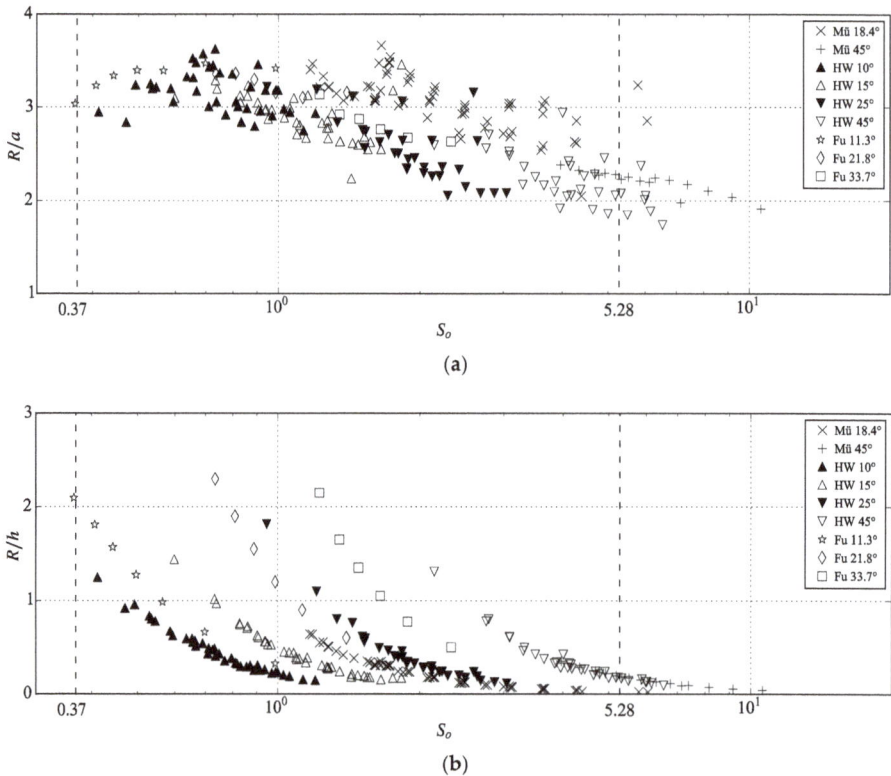

(a)

(b)

Figure 3. (a) Runup height over wave amplitude R/a and (b) runup height over stillwater depth R/h versus slope parameter S_o for experiments from data sources from Müller (Mü) [7] (without $\beta = 90°$), Hall and Watts (HW) [14], and Fuchs (Fu) [20].

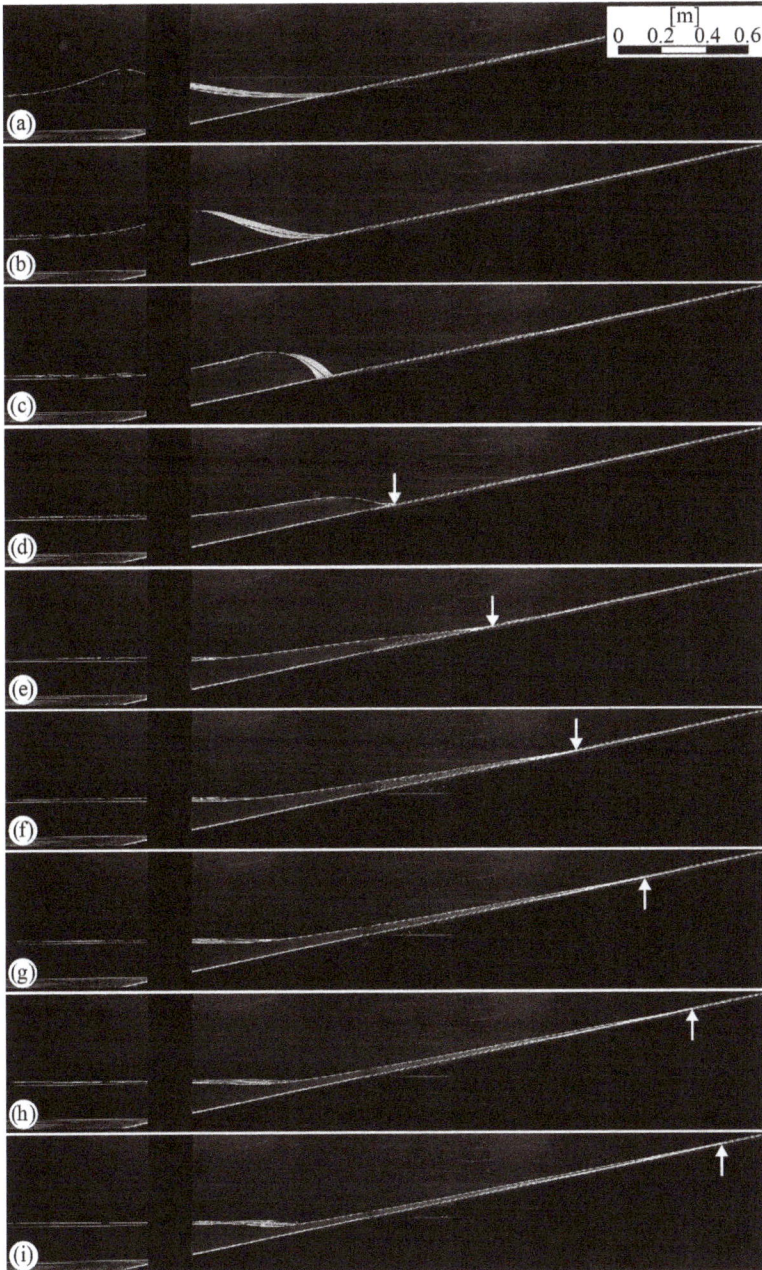

Figure 4. Solitary wave runup for tanβ = 1/5.0 ($\beta \approx$ 11.3°), ε = 0.69, and S_o = 0.37, time increment between images Δt = 0.2 s, flow front indicated by arrow (reproduced from Fuchs [20]).

4. Results

4.1. Existing Prediction Equations

Equation (1) by Müller [7] predicts its underlying data well within a ±20% scatter range (Figure 5). Only few experiments with β = 18.4° exceed this range, i.e., the actually measured runup heights are more than 20% larger than the predicted values. Equation (2) was applied to get the effective wave length L_{Sol} to allow for the prediction of the solitary wave experiments by Hall and Watts [14], Fuchs [20], Street and Camfield [21], and Maxworthy [19]. However, these experiments are broadly underestimated and the measured runup heights are up to 70% higher than their prediction.

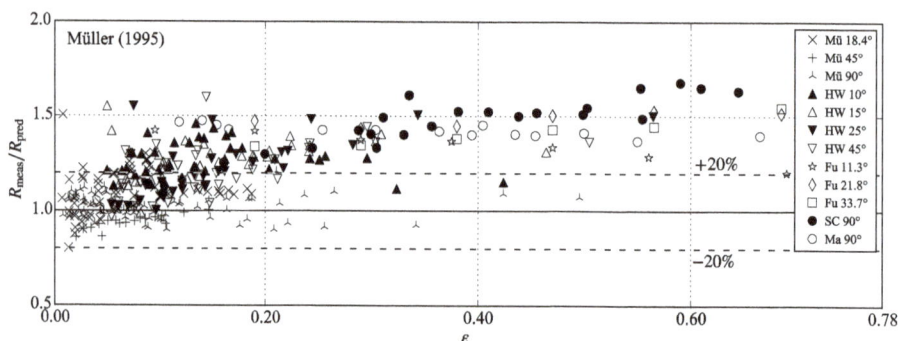

Figure 5. Measured over predicted (Equation (1) by Müller [7]) runup height R_{meas}/R_{pred} versus relative wave crest amplitude ε for experiments from data sources from Müller (Mü) [7], Hall and Watts (HW) [14], Fuchs (Fu) [20], Street and Camfield (SC) [21], and Maxworthy (Ma) [19].

The data by Hall and Watts [14] is predicted well within a ±20% scatter range by their empirically derived Equation (3) (Figure 6). Additionally, the runup experiments by Street and Camfield [21] and Maxworthy [19] with β = 90° are predicted within this range, although vertical walls were not considered by Hall and Watts [14]. For small relative wave amplitudes ε < 0.05 to 0.1, the measured runup heights are up to 50% larger than the predicted values. Mainly the impulse wave data by Müller [7] is affected by this underestimation, as it features small ε. For ε > 0.1, impulse wave runup is also well predicted.

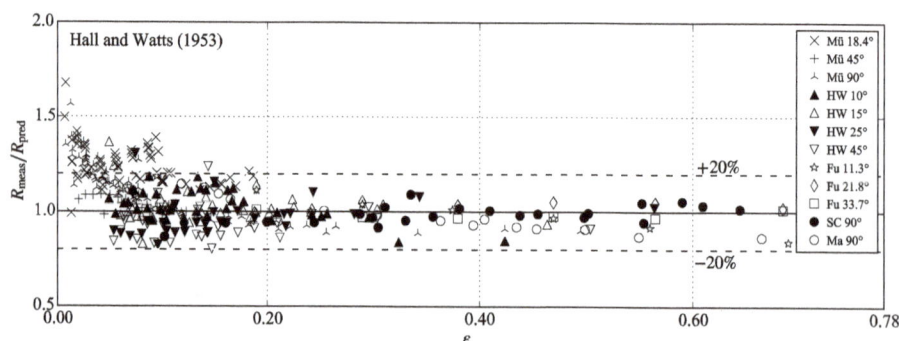

Figure 6. Measured over predicted (Equation (3) by Hall and Watts [14]) runup height R_{meas}/R_{pred} versus relative wave crest amplitude ε for experiments from data sources from Müller (Mü) [7], Hall and Watts (HW) [14], Fuchs (Fu) [20], Street and Camfield (SC) [21], and Maxworthy (Ma) [19].

Figure 7 excludes runup data at vertical walls, since Equation (4) by Fuchs and Hager [15] is not defined for $\beta = 90°$. Equation (4) predicts its underlying data well. Additionally, the measured runup heights at $\beta = 10°$, $15°$, and $18.4°$ scatter within $\pm 20\%$ of the prediction. However, the runup data for $\beta = 25°$ and $45°$ are predicted too conservatively for $\varepsilon < 0.2$, i.e., the actual runup heights are up to 40% smaller than their predicted values.

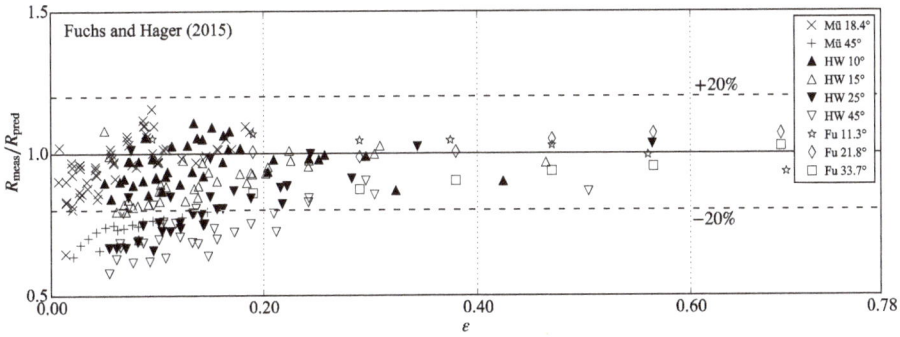

Figure 7. Measured over predicted (Equation (4) by Fuchs and Hager [15]) runup height R_{meas}/R_{pred} versus relative wave crest amplitude ε for experiments from data sources from Müller (Mü) [7] (without $\beta = 90°$), Hall and Watts (HW) [14], and Fuchs (Fu) [20].

Equation (5) by Synolakis [16] yields runup heights equal to zero for $\beta = 90°$. Therefore, these experiments are excluded from Figure 8. The measured values scatter broadly around the predicted runup heights both for solitary and impulse waves. While the measured runup heights for $\varepsilon > 0.2$ are up to 50% smaller, the measurements for $\varepsilon < 0.2$ are up to 100% above the prediction. The slope parameter S_0 of the latter experiments is close to 5.28, the upper limiting criterion of Equation (5).

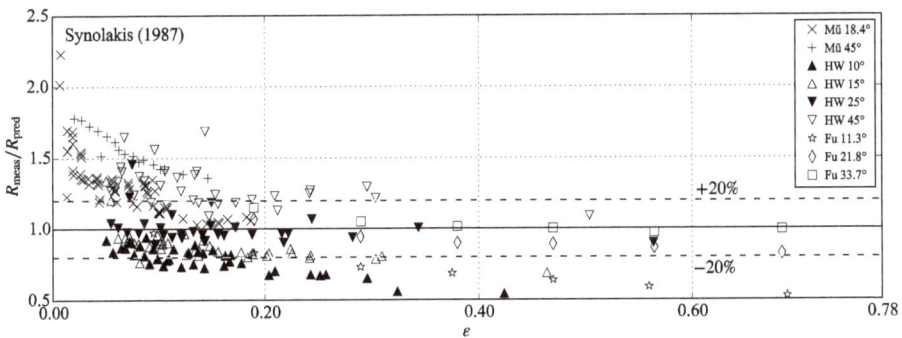

Figure 8. Measured over predicted (Equation (5) by Synolakis [16]) runup height R_{meas}/R_{pred} versus relative wave crest amplitude ε for experiments from data sources from Müller (Mü) [7] (without $\beta = 90°$), Hall and Watts (HW) [14], and Fuchs (Fu) [20].

Equation (6) by Su and Mirie [17] was analytically derived for solitary wave runup at a vertical wall. Both impulse and solitary wave experiments with $\beta = 90°$ scatter narrowly within $\pm 20\%$ around the prediction (Figure 9). There is no distinct effect of ε on the prediction quality, even for small ε. Also, the data with $\beta = 45°$ scatter within this range. With decreasing slope angle β, the measured runup heights are more and more underestimated. For $\beta < 20°$, the measurements are up to 80% larger than the predictions. However, this underestimation appears to be smaller for large ε.

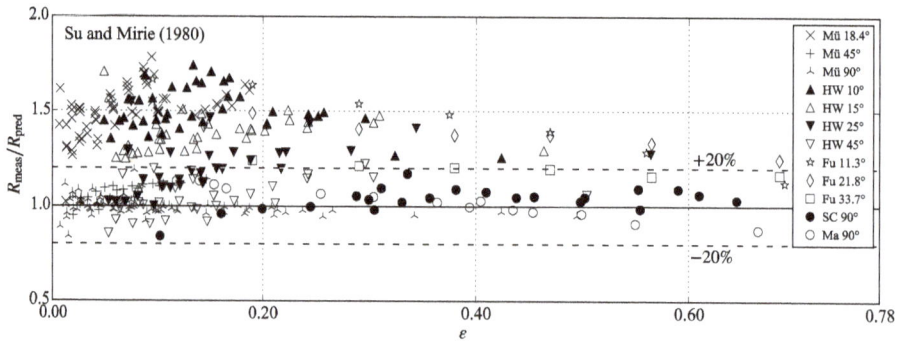

Figure 9. Measured over predicted (Equation (6) by Su and Mirie [17]) runup height $R_{\mathrm{meas}}/R_{\mathrm{pred}}$ versus relative wave crest amplitude ε for experiments from data sources from Müller (Mü) [7], Hall and Watts (HW) [14], Fuchs (Fu) [20], Street and Camfield (SC) [21], and Maxworthy (Ma) [19].

4.2. New Prediction Equation

Based on the findings in the previous sections, a new prediction approach is proposed. While Equation (1) by Müller [7] adequately captures the effect of the slope angle β from 18.4° to 90°, Equation (6) by Su and Mirie [17] yields good runup predictions at vertical walls for a broad range of wave amplitudes ε. In Figure 10, the runup heights of the overall experimental dataset are approximated ($R^2 = 0.98$) with:

$$\frac{R}{h} = \left(2\varepsilon + \frac{1}{2}\varepsilon^2 + \frac{3}{4}\varepsilon^3\right)\left(\frac{90°}{\beta}\right)^{0.2}.\tag{9}$$

The first bracket includes Equation (6) by Su and Mirie [17] and the second bracket includes the effect of β from Equation (1) by Müller [7]. For $\beta = 90°$ the latter term equals to 1 and for $\beta = 10°$ it is approximately 1.55, i.e., the runup height increases with decreasing β. The measured runup heights scatter within circa ±25% of the prediction (Figure 10). The 2.5th and 97.5th percentile whiskers in Figure 11a show that more than 95% of the experiments are within a ±20% range. The maximum deviations of single experiments are within ±30%. These experiments include $\varepsilon < 0.2$ as well as large ε at comparatively gentle slope angles $\beta = 10°$ and 11.3°. Figure 11b shows the direct comparison of the measured relative runup heights R_{meas}/h versus the predicted R_{pred}/h.

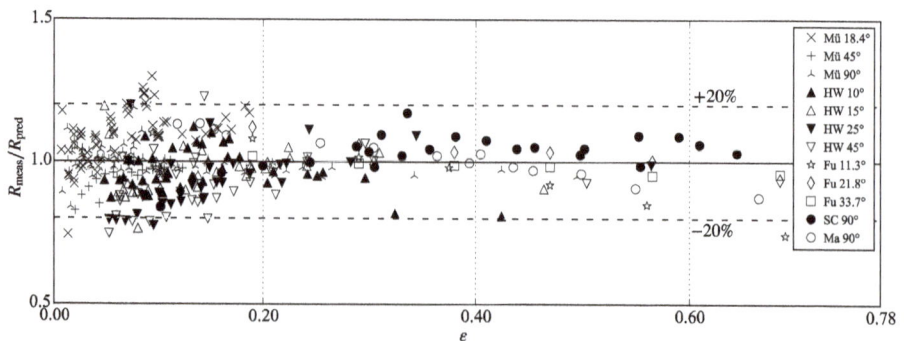

Figure 10. Measured over predicted (Equation (9)) runup height $R_{\mathrm{meas}}/R_{\mathrm{pred}}$ versus relative wave crest amplitude ε for experiments from data sources from Müller (Mü) [7], Hall and Watts (HW) [14], Fuchs (Fu) [20], Street and Camfield (SC) [21], and Maxworthy (Ma) [19].

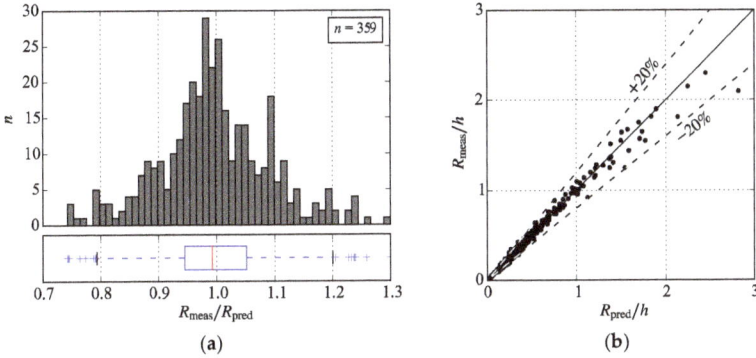

Figure 11. (a) Histogram and boxplot with whiskers at 2.5th and 97.5th percentile of measured over predicted (Equation (9)) runup heights R_{meas}/R_{pred} and (b) measured versus predicted runup heights R over stillwater depth h of the overall dataset.

As a simplified version of Equation (9), the following prediction equation for non-breaking impulse and solitary wave runup on steep to vertical slopes is proposed ($R^2 = 0.98$):

$$\frac{R}{h} = 2\varepsilon e^{0.4\varepsilon}\left(\frac{90°}{\beta}\right)^{0.2}.$$ (10)

The term $2\varepsilon e^{0.4\varepsilon}$ in Equation (10) approximates Su and Mirie's [17] Equation (6) by multiplying the minimum runup height of 2ε for very small relative wave amplitudes ε with an exponential function accounting for the second- and third-order effects, which become significant for large ε. 33 additional experiments are added as a validation dataset to Figure 12, including data from Pedersen et al. [27] ($n = 5$; $\beta = 10°$), Li and Raichlen [28] ($n = 22$; $\beta = 25.7°$), and Losada et al. [29] ($n = 6$; $\beta = 45°, 70°, 90°$; taken from Maiti and Sen [30]). The additional relative wave crest amplitudes ε range from 0.026 to 0.48. The scatter plot ($R^2 = 0.98$) of Equation (10) in Figure 12 is very similar to Figure 10 based on Equation (9). As shown in Figure 13a, the overall scatter is slightly shifted to the side of caution, i.e., overestimation, by this approximation therefore having a minor effect on the overall prediction quality. Figure 13b compares the approximation $2\varepsilon e^{0.4\varepsilon}$ with Equation (6). The deviations are around ±2% within the range of the analyzed dataset ($\varepsilon \leq 0.69$) and −4% for a maximum $\varepsilon = 0.78$ (McCowan [31]).

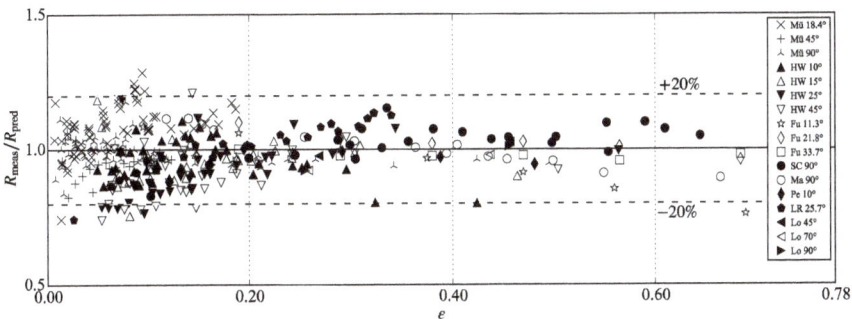

Figure 12. Measured over predicted (Equation (10)) runup height R_{meas}/R_{pred} versus relative wave crest amplitude ε for experiments from data sources from Müller (Mü) [7], Hall and Watts (HW) [14], Fuchs (Fu) [20], Street and Camfield (SC) [21], Maxworthy (Ma) [19], Pedersen et al. (Pe) [27], Li and Raichlen (LR) [28], and Losada et al. (Lo) [29].

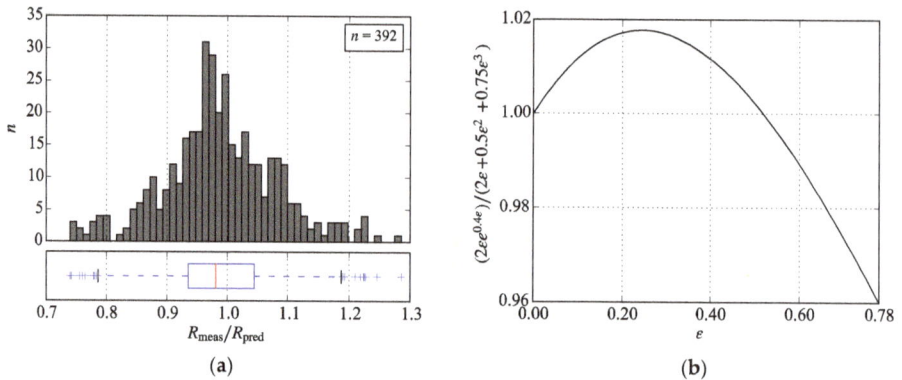

Figure 13. (a) Histogram and boxplot with whiskers at 2.5th and 97.5th percentile of measured over predicted (Equation (10)) runup height R_{meas}/R_{pred} of the overall dataset; (b) $2\varepsilon e^{0.4\varepsilon}$ over Equation (6) versus relative wave amplitude ε.

5. Discussion

With $S_o \geq 0.37$, the overall dataset is assumed to include non-breaking wave runup. Considering a maximum relative wave crest amplitude $\varepsilon = 0.78$ before runup according to McCowan [31], the minimum slope angle satisfying $S_o = 0.37$ is $\beta = 12°$. For this study, also experiments with $\beta = 10°$ and 11.3° were analyzed, which would lead to wave breaking for $\varepsilon = 0.78$ according to S_o. Equation (10) tends to overestimate runup heights by large ε on these slopes (Figure 12). Therefore $S_o = 0.37$ is considered the lower application boundary of Equation (10) for β from 10° to 90°. Pujara et al. [10] proposed a more conservative breaking criterion $S_o \approx 0.4$ to 0.5. With $S_o > 0.5$ to ensure no wave-breaking at $\beta = 10°$, the maximum amplitude is $\varepsilon \leq 0.3$, which still provides a useful range of application for less steep slopes.

Compared to Equation (1) by Müller [7], the new prediction Equation (10) contains solely the relative wave crest amplitude ε as the governing wave parameter instead of the wave height H and the wave length L. For the assessment of landslide generated impulse wave events, prediction equations are applied in sequence to cover a particular process chain, e.g., wave generation and runup (Bregoli et al. [8], McFall and Fritz [9]). In this context, the maximum relative scatter of a target value, e.g., R, is derived from the scatter of its individual input parameters, e.g., ε, H and/or L (Heller et al. [1]). While the scatter is not significantly altered by substituting H and L with ε for predicting Müller's [7] measurements (Figures 5 and 12), the prediction uncertainty for the entire process chain may be reduced by including fewer parameters.

Fuchs and Hager [32] conducted scale family experiments of solitary wave runup and observed no significant scale effects for $h \geq 0.08$ m at $\beta = 11.3°$. For their smallest investigated $\varepsilon = 0.1$, this corresponds to a minimum runup height $R = 25$ mm. The dataset by Müller [7] contains six experiments at $\beta = 18.4°$ with R between 12 and 24 mm induced by ε from 0.007 to 0.014. As shown in Figure 12, these experiments scatter around ±20% of the prediction. This range reflects the measurement accuracy as well as the experimental accuracy from repeatability tests (Müller [7]). As no distinct trend is observed in the data, scale effects are considered negligible. Also, the experiments by Maxworthy [19] feature $R < 25$ mm. However, these experiments were conducted at a much steeper slope with $\beta = 90°$. In addition, the measured runup heights show no distinct deviation compared to the experiments by Street and Camfield [21], which were conducted at a three to four times larger scale (Figure 9).

The experimental setups considered in this study feature two-dimensional, plane, and smooth runup slopes, which represent a simplification of shorelines at prototype scale. Additional prototype parameters include three-dimensional slope features, non-constant slopes, and rough surfaces. Strong curvatures along the shoreline cause flow diversion and concentration, respectively, which might lead to a significant over- or underestimation of the actual runup height. Therefore, Equation (10) should only be applied to evenly formed slope bathymetries and topographies. Non-constant slopes complicate the determination of a single slope angle β. A sensitivity analysis allows for assessing the influence of β for the slope range derived from field data. The term of Equation (10) including the slope angle β yields larger runup heights for decreasing β. A decrease from 90° to 45° leads to an increase in runup height of 15%. A decrease from 20° to 10° also leads to an increase of 15%. Therefore, the effect of β is stronger for lower slope angles. Teng et al. [33] conducted experiments on solitary wave runup on both smooth and rough slopes. The roughness effect was found to be negligible on relatively steep slopes ($\beta \geq 20°$), while it reduced the measured runup heights by up to approximately 30% for $\beta = 15°$ and by 50% for $\beta = 10°$. However, runup height estimation based on Equation (10), which is derived from experiments featuring smooth slopes, would err on the side of caution. Finally, the scatter range of $\pm 20\%$ needs to be taken into account as a safety margin for runup height predictions at prototype scale.

6. Conclusions

The runup height of impulse waves as well as solitary waves on slope angles between 10° and 90° was analyzed. While a solitary wave features a single wave crest without a following trough, an impulse wave train consists of multiple wave crests and troughs. A dataset with $n = 359$ experiments was compiled from literature including runup heights by both wave types. Regarding impulse wave runup, the first wave within the wave train was analyzed based on the experiments by Müller [7]. The overall dataset was compared with both empirical and analytical equations from literature, and a new semi-empirical prediction equation was presented. Additionally, $n = 33$ experiments were included as a validation dataset. The main findings are:

- Solitary waves are a suitable proxy for modelling the runup height of the first wave within an impulse wave train;
- Commonly applied prediction equations may significantly underestimate the measured runup heights by small relative wave crest amplitudes $\varepsilon < 0.1$;
- A new equation for non-breaking impulse wave runup on slope angles from 10° to 90° predicts the overall dataset within a $\pm 20\%$ scatter for ε between 0.007 and 0.69 as the single wave input parameter.

Author Contributions: Data analysis, F.M.E; writing—original draft preparation and editing, F.M.E; writing—review, R.M.B.; funding acquisition, R.M.B.

Funding: This work was funded by the Swiss Federal Office of Energy SFOE / Bundesamt für Energie BFE (Project No. SI/501802-01) and is part of the Swiss Competence Center for Energy Research—Supply of Electricity (SCCER-SoE).

Acknowledgments: The authors would like to thank the experimenters for publishing their data in a way allowing for reuse. Willi H. Hager is acknowledged for helpful discussions. Figure 4 was provided by courtesy of Helge Fuchs.

Conflicts of Interest: The authors declare no conflict of interest.

Nomenclature

Symbol	Term	Unit
a	Wave crest amplitude	[m]
c	Wave crest celerity	[m/s]
c_{Sol}	Solitary wave crest celerity	[m/s]
h	Stillwater depth	[m]

H	Wave height	[m]
g	Gravitational acceleration	[m/s^2]
L	Wave length	[m]
L_{Sol}	Solitary wave length	[m]
n	Number of experiments	[-]
R	Runup height	[m]
R^2	Coefficient of determination	[-]
S_o	Slope parameter	[-]
β	Slope angle	[°]
ε	Relative wave crest amplitude	[-]
Fu	Fuchs [20]	
HW	Hall and Watts [14]	
Li	Li and Raichlen [28]	
Lo	Losada et al. [29]	
Ma	Maxworthy [19]	
Mü	Müller [7]	
Pe	Pedersen et al. [27]	
SC	Street and Camfield [21]	

References

1. Heller, V.; Hager, W.H.; Minor, H.-E. Landslide generated impulse waves in reservoirs: Basics and computation. In *VAW-Mitteilung 211*; Boes, R., Ed.; ETH Zurich: Zürich, Switzerland, 2009.
2. Roberts, N.J.; McKillop, R.; Hermanns, R.L.; Clague, J.J.; Oppikofer, T. Preliminary global catalogue of displacement waves from subaerial landslides. In *Landslide Science for a Safer Geoenvironment*; Sassa, K., Canuti, P., Yin, Y., Eds.; Springer: Cham, Switzerland, 2014; Volume 3, pp. 687–692. ISBN 978-3-319-04996-0.
3. Miller, D.J. *Giant Waves in Lituya Bay, Alaska*; Geological Survey Professional Paper No. 354-C; U.S. Government Printing Office: Washington, DC, USA, 1960.
4. Roberts, N.J.; McKillop, R.J.; Lawrence, M.S.; Psutka, J.F.; Clague, J.J.; Brideau, M.-A.; Ward, B.C. Impacts of the 2007 landslide-generated tsunami in Chehalis Lake, Canada. In *Landslide Science and Practice*; Margottini, C., Canuti, P., Sassa, K., Eds.; Springer: Berlin/Heidelberg, Germany, 2013; Volume 6, pp. 133–140. ISBN 978-3-642-31319-6.
5. Higman, B.; Shugar, D.H.; Stark, C.P.; Ekström, G.; Koppes, M.N.; Lynett, P.; Dufresne, A.; Haeussler, P.J.; Geertsema, M.; Gulick, S.; et al. The 2015 landslide and tsunami in Taan Fiord, Alaska. *Sci. Rep. UK* **2018**, *8*, 12993. [CrossRef]
6. Fuchs, H.; Boes, R. Berechnung felsrutschinduzierter Impulswellen im Vierwaldstättersee. *Wasser Energ. Luft* **2010**, *102*, 215–221. (In German) [CrossRef]
7. Müller, D.R. Auflaufen und Überschwappen von Impulswellen an Talsperren. In *VAW-Mitteilung 137*; Vischer, D., Ed.; ETH Zurich: Zürich, Switzerland, 1995. (In German)
8. Bregoli, F.; Bateman, A.; Medina, V. Tsunamis generated by fast granular landslides: 3D experiments and empirical predictors. *J. Hydraul. Res.* **2017**, *55*, 743–758. [CrossRef]
9. McFall, B.C.; Fritz, H.M. Runup of granular landslide-generated tsunamis on planar coasts and conical islands. *J. Geophys. Res. Oceans* **2017**, *122*, 6901–6922. [CrossRef]
10. Pujara, N.; Liu, P.L.-F.; Yeh, H. The swash of solitary waves on a plane beach: Flow evolution, bed shear stress and run-up. *J. Fluid Mech.* **2015**, *779*, 556–597. [CrossRef]
11. Hafsteinsson, H.J.; Evers, F.M.; Hager, W.H. Solitary wave run-up: Wave breaking and bore propagation. *J. Hydraul. Res.* **2017**, *55*, 787–798. [CrossRef]
12. Dean, R.G.; Dalrymple, R.A. *Water Wave Mechanics for Engineers and Scientists*; World Scientific Publishing: Singapore, 1991.
13. Lo, H.-Y.; Park, Y.S.; Liu, P.L.-F. On the run-up and back-wash processes of single and double solitary waves—An experimental study. *Coast. Eng.* **2013**, *80*, 1–14. [CrossRef]

14. Hall, J.V.; Watts, G.M. *Laboratory Investigation of the Vertical Rise of Solitary Wave on Impermeable Slopes*; Technical Memo Report No. 33; U.S. Army Corps of Engineers, Beach Erosion Board: Washington, DC, USA, 1953.
15. Fuchs, H.; Hager, W.H. Solitary impulse wave transformation to overland flow. *J. Waterw. Port C ASCE* **2015**, *141*, 04015004. [CrossRef]
16. Synolakis, C.E. The runup of solitary waves. *J. Fluid Mech.* **1987**, *185*, 523–545. [CrossRef]
17. Su, C.H.; Mirie, R.M. On head-on collisions between two solitary waves. *J. Fluid Mech.* **1980**, *98*, 509–525. [CrossRef]
18. Cooker, M.J.; Weidman, P.D.; Bale, D.S. Reflection of a high-amplitude solitary wave at a vertical wall. *J. Fluid Mech.* **1997**, *342*, 141–158. [CrossRef]
19. Maxworthy, T. Experiments on collisions between solitary waves. *J. Fluid Mech.* **1976**, *76*, 177–186. [CrossRef]
20. Fuchs, H. Solitary impulse wave run-up and overland flow. In *VAW-Mitteilung 221*; Boes, R., Ed.; ETH Zurich: Zürich, Switzerland, 2013.
21. Street, R.L.; Camfield, F.E. Observations and experiments on solitary wave deformation. In Proceedings of the 10th International Conference on Coastal Engineering, Tokyo, Japan, September 1966; ASCE: Reston, VA, USA, 1966. [CrossRef]
22. Russell, J.S. *Report on Waves*; Report of the 14th Meeting of the British Association for the Advancement of Science; British Association: York, UK, 1944; pp. 311–390.
23. McFall, B.C.; Fritz, H.M. Physical modelling of tsunamis generated by three-dimensional deformable granular landslides on planar and conical island slopes. *Proc. R. Soc. A Math. Phys.* **2016**, *472*, 20160052. [CrossRef] [PubMed]
24. Evers, F.M.; Hager, W.H.; Boes, R.M. Spatial impulse wave generation and propagation. *J. Waterw. Port C ASCE* **2019**, in press. [CrossRef]
25. Rohatgi, A.; Rehberg, S.; Stanojevic, Z. ankitrohatgi/WebPlotDigitizer: Version 4.1 of WebPlotDigitizer (Version v4.1). *Zenodo* **2018**. [CrossRef]
26. Grilli, S.T.; Svendsen, I.A.; Subramanya, R. Breaking criterion and characteristics for solitary waves on slopes. *J. Waterw. Port C ASCE* **1997**, *123*, 102–112. [CrossRef]
27. Pedersen, G.K.; Lindstrøm, E.; Bertelsen, A.F.; Jensen, A.; Laskovski, D.; Sælevik, G. Runup and boundary layers on sloping beaches. *Phys. Fluids* **2013**, *25*, 012102. [CrossRef]
28. Li, Y.; Raichlen, F. Solitary wave runup on plane slopes. *J. Waterw. Port C ASCE* **2001**, *127*, 33–44. [CrossRef]
29. Losada, M.A.; Vidal, C.; Nunez, J. *Sobre El Comportamiento de Ondas Propagádose por Perfiles de Playa en Barra y Diques Sumergidos*; Dirección General de Puertos y Costas Programa de Clima Marítimo; Universidad de Cantabria: Cantabria, Spain, 1986. Publicación No. 16. (In Spanish)
30. Maiti, S.; Sen, D. Computation of solitary waves during propagation and runup on a slope. *Ocean Eng.* **1999**, *26*, 1063–1083. [CrossRef]
31. McCowan, J. On the highest wave of permanent type. *Lond. Edinb. Dublin Philos. Mag. J. Sci.* **1894**, *38*, 351–358. [CrossRef]
32. Fuchs, H.; Hager, W.H. Scale effects of impulse wave run-up and run-over. *J. Waterw. Port C ASCE* **2013**, *138*, 303–311. [CrossRef]
33. Teng, M.H.; Feng, K.; Liao, T.I. Experimental study on long wave run-up on plane beaches. In Proceedings of the 10th International Offshore and Polar Engineering Conference, Seattle, WA, USA, 28 May–2 June 2000; The International Society of Offshore and Polar Engineers, ISOPE-I-00-310. pp. 660–664.

© 2019 by the authors. Licensee MDPI, Basel, Switzerland. This article is an open access article distributed under the terms and conditions of the Creative Commons Attribution (CC BY) license (http://creativecommons.org/licenses/by/4.0/).

Journal of
Marine Science and Engineering

MDPI

Article

Deciphering the Tsunami Wave Impact and Associated Connection Forces in Open-Girder Coastal Bridges

Denis Istrati [1,*], Ian Buckle [1], Pedro Lomonaco [2] and Solomon Yim [3]

[1] Department of Civil and Environmental Engineering, University of Nevada, Reno, 1664 N. Virginia Street, Reno, NV 89557-0258, USA; igbuckle@unr.edu

[2] O.H. Hinsdale Wave Research Laboratory, Oregon State University, School of Civil and Construction Engineering, 3550 SW Jefferson Way, Corvallis, OR 97331, USA; pedro.lomonaco@oregonstate.edu

[3] School of Civil and Construction Engineering, Oregon State University, 350 Owen Hall, Corvallis, OR 97331, USA; solomon.yim@oregonstate.edu

* Correspondence: distrati@unr.edu; Tel.: +1-775-784-6937

Received: 9 October 2018; Accepted: 29 November 2018; Published: 5 December 2018

Abstract: In view of the widespread damage to coastal bridges during recent tsunamis (2004 Indian Ocean and 2011 in Japan) large-scale hydrodynamic experiments of tsunami wave impact on a bridge with open girders were conducted in the Large Wave Flume at Oregon State University. The main objective was to decipher the tsunami overtopping process and associated demand on the bridge and its structural components. As described in this paper, a comprehensive analysis of the experimental data revealed that: (a) tsunami bores introduce significant slamming forces, both horizontal (Fh) and uplift (Fv), during impact on the offshore girder and overhang; these can govern the uplift demand in connections; (b) maxFh and maxFv do not always occur at the same time and contrary to recommended practice the simultaneous application of maxFh and maxFv at the center of gravity of the deck does not yield conservative estimates of the uplift demand in individual connections; (c) the offshore connections have to withstand the largest percentage of the total induced deck uplift among all connections; this can reach 91% and 124% of maxFv for bearings and columns respectively, a finding that could explain the damage sustained by these connections and one that has not been recognized to date; (e) the generation of a significant overturning moment (OTM) at the initial impact when the slamming forces are maximized, which is the main reason for the increased uplift in the offshore connections; and (f) neither maxFv nor maxOTM coincide always with the maximum demand in each connection, suggesting the need to consider multiple combinations of forces with corresponding moments or with corresponding locations of application in order to identify the governing scenario for each structural component. In addition the paper presents "tsunami demand diagrams", which are 2D envelopes of (Fh, Fv) and (OTM, Fv) and 3D envelopes of (Fh, Fv, OTM), as visual representations of the complex variation of the tsunami loading. Furthermore, the paper reveals the existence of a complex bridge inundation mechanism that consists of three uplift phases and one downward phase, with each phase maximizing the demand in different structural components. It then develops a new physics-based methodology consisting of three load cases, which can be used by practicing engineers for the tsunami design of bridge connections, steel bearings and columns. The findings in this paper suggest the need for a paradigm shift in the assessment of tsunami risk to coastal bridges to include not just the estimation of total tsunami load on a bridge but also the distribution of this load to individual structural components that are necessary for the survival of the bridge.

Keywords: tsunami; experiments; wave impact; bore; solitary wave; slamming force; bridge; deck; connections; bearings

1. Introduction

In the last two decades, large magnitude earthquakes with epicenters in the ocean (Indian Ocean 2004, Chile 2010 and Japan 2011) have generated tsunami waves of significant heights that caused unprecedented damage to coastal communities. Ports, buildings and infrastructure were severely damaged and bridges were washed away, cutting lifelines and hindering the efforts of rescue teams to provide help to the people in need. In the 2004 Indian Ocean Tsunami, 81 bridges located on the coast of Sumatra were washed away [1]. In the 2011 Great East Japan Earthquake in Japan, many bridges were able to withstand the strong shaking; however, 252 bridges were washed away or moved by the tsunami [2]. The most severe and common type of failure in these bridges was the breaking of the connections between the superstructure and the substructure, which resulted in the unseating and wash out of the bridge deck by the tsunami waves (Figure 1).

Figure 1. Damaged bridges after the 2011 Great East Japan Earthquake: (**a**) Koizumi bridge on the left (photo: E. Monzon) and (**b**) Utatsu Bridge on the right (photo: I. G. Buckle).

These unforeseen events demonstrated the vulnerability of bridges to tsunami waves and highlighted the need to understand the tsunami-induced loading. In response to this need several studies have been published in recent years, including on-site surveys and damage analysis [3–5], small-scale experiments in wave flumes [6–9] and numerical simulations [10–17]. On-site investigations conducted by various research teams analyzed the failed bridges and revealed that the overflow can occur either in the form of transverse drag due to large horizontal wave forces or in the form of uplift and overturning due to the combination of large vertical and horizontal tsunami forces [5,18,19].

Some of the experimental studies investigated tsunami loads on flat slabs [8], decks with girders [6,9,20,21] and box-shaped decks [7,22,23]. Several studies simulated the tsunami waves via unbroken but transformed solitary waves, while others via turbulent bores. In most of these experiments, the researchers constructed their bridge models from acrylic, wood or steel, supported the deck rigidly and measured both pressures and total forces. Furthermore, they were all small-scale experiments with scale factors ranging between 1:100 and 1:35. Hoshikuma et al. [24] conducted experiments to study the tsunami effects on bridges at a larger scale, equal to 1:20. They examined several different cross-sections with deck specimens made of acrylic or wood and connected rigidly to a pier at the center of the superstructure. To the authors' best knowledge these experiments were the first ones to measure the tsunami demand on the connections between the superstructure and substructure and demonstrated that the offshore bearings were uplifted while the onshore were compressed implying the existence of overturning moment. Despite the fact that all these small and medium-scale experiments gave an insight into the tsunami forces on bridges, they might have significant scale effects because: (a) it is not possible to scale the atmospheric pressure, a fact that will affect the interaction of the wave with the trapped air between the girders and possibly modify the applied pressures and forces; and (b) at such scales it is not easy to accurately simulate the structural properties [25].

Apart from the experimental studies, several numerical analyses have been conducted to study the tsunami impact on bridges. Among others, Lau et al. [6] and Bricker and Nakayama [14] conducted computational fluid dynamics (CFD) analyses and obtained the total applied tsunami loading. In the first case, the researchers used the CFD results for the development of predictive loading equations, while in the latter case, the calculated tsunami loading was directly compared with the capacity of the Utatsu Bridge in Japan revealing that the bridge failed due to the deck superelevation, nearby structures and trapped air. Another study conducted numerical analyses of a rigid bridge model (Azadbakht and Yim [15]) with a finite element method (FEM)-based multi-physics commercial software program called LS-DYNA and developed by the Livermore Software Technology Corporation (LSTC), Livermore, California, for the development of a tsunami load estimation method. Murakami et al. [26] calculated the pressures on a rigid bridge model via CFD analyses and then used them as external loads on a slab-type bridge model with flexible springs in order to get an estimate of the uplift demand in the individual bearings. Istrati and Buckle [16] conducted advanced fluid-structure interaction (FSI) analyses in LS-DYNA (V971 R6.1.2) using an equivalent 2D bridge model with flexible deck and connections, which showed that the dynamic characteristics of the bridge affected both the tsunami load applied on the bridge, as well as the forces in the connections. In addition, the study showed the existence of a rotational mode during the impact of tsunami waves, which put the offshore bearings in tension and the onshore ones in compression, increasing consequently the demand on the offshore connections.

More recently Motley et al. [27] developed 2D and 3D CFD numerical models of a 1:20 scale bridge model in OpenFOAM to examine the effect of the bridge skewness. Several skew angles between 0 and 40 degrees were examined and it was revealed that the skew bridge is subjected to pitching and spinning moments and that there exists a force normal to the abutments that could lead to unseating. Another, recent study conducted by Wei and Darlymple [28], simulated the same 1:20 scale straight bridge as the previous researchers, using the weakly compressible smoothed particle hydrodynamics (SPH) method in GPUSPH and demonstrated the possibility of mitigating the tsunami effects on bridges via the use of an offshore breakwater or the existence of another bridge on the seaward side of the main bridge. Recently, Zhu et al. [29] implemented the particle finite element method (PFEM) in OpenSees to simulate tsunami impact on bridge decks and validated the methodology again using the small-scale experiments conducted in [24]. Although most of the aforementioned studies focused on the total applied wave loading, more recent studies ([30–32]) examined the tsunami-induced loading on individual girders and/or deck chambers, and revealed that the upstream girder and the upstream part of the deck have to withstand a large percentage of the total loading, which suggests the need to investigate the tsunami-induced local effects on individual bridge components (e.g., rails, girders, deck chambers). It must be noted that several of the studies conducted in the past have focused on understanding the hurricane induced forces on bridges [33–35] and some of their findings could possibly contribute to the understanding of the tsunami wave impact on bridges as well; however, due to limited space these studies are not reviewed herein. The readers could find a comprehensive review of the hurricane induced bridge forces in [36].

As described previously most of the studies to date investigated the total induced tsunami wave loading, a few recent ones focused on the tsunami demand on individual girders and deck chambers and only very few studies considered the demand on the bearings and connections. Given the fact that (a) the latter studies revealed significant demand on the offshore bearings and connections and (b) the breaking of the bearing connections was the main type of bridge damage witnessed in recent tsunamis, it becomes critical to quantify the tsunami demand on these components and decipher the underlying physics. Moreover, given the fact that most experimental studies were conducted at a small-scale with possibly significant scale-effects, while most of the numerical studies were 2D CFD analyses that did not consider the structural properties, it is important to conduct large-scale experiments of tsunami wave impact on bridge specimens with realistic dynamic properties (flexibility and inertia), which will not be subjected to the aforementioned limitations. Therefore, this paper will present a comprehensive analysis of 1:5 scale hydrodynamic experiments of tsunami impact on a representative coastal bridge with open girders conducted at Oregon State University, with particular focus on the spatial and temporal variation of both the total applied loading and the uplift demand in the bearings, connections and columns. The study will also present experimental evidence that decipher the tsunami inundation mechanism and a physics-based simple methodology that can be used by practicing engineers for estimating the tsunami-induced uplift loading in individual connections.

2. Description of Hydrodynamic Experiments

2.1. Experimental Setup

To gain a realistic insight into the tsunami effects on coastal bridge decks, large-scale hydrodynamic experiments were conducted in the Large Wave Flume (LWF) at the O.H. Hinsdale Wave Research Laboratory (HWRL) at Oregon State University. The flume is 104.24 m long, 3.66 m wide, and 4.57 m deep. The maximum depth for tsunami-type wave generation is 2 m, and the maximum wave height for this depth is 1.40 m. The LWF is equipped with a piston-type dry-back wavemaker with a 4.2 m maximum stroke hydraulic actuator assembly and has a movable/adjustable bathymetry made of 20 square configurable concrete slabs. The flume includes a series of bolt-holes vertical patterns every 3.66 m along the flume for supporting test specimens and for creating different bathymetries. In order to identify the bathymetry that would permit the testing of both unbroken solitary waves and bores, and determine the optimum location of the bridge in the flume, parametric CFD analyses of the whole flume were conducted prior to the experiments. These analyses revealed that a slope of 1:12 at the beginning, followed by a horizontal bathymetry 40.2 m long and another 1:12 slope at the end of the flume was the most appropriate (Figure 2). In addition, the optimum location for the bridge was between bays 14 and 15, at a distance of 58.8 m from the wavemaker, in order to allow for the bore to form and overtop the bridge.

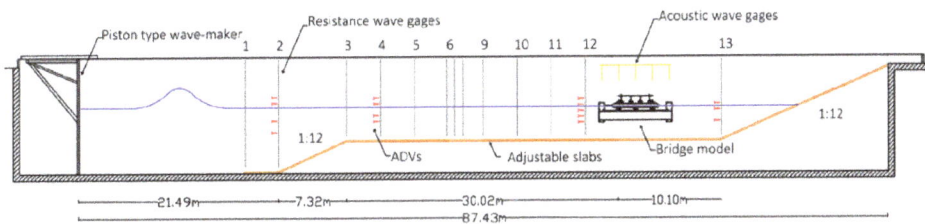

Figure 2. Cross-section of the Large Wave Flume (LWF) depicting the bathymetry, bridge location and flume instrumentation (not to scale).

For the hydraulic experiments a composite bridge model with four I-girders was designed and constructed at a 1:5 scale. The in-plane dimensions of the bridge deck were 3.45 m length and 1.94 m width. As shown in Figure 3 the steel girders were connected with cross-frames at the end supports. Additional cross-frames had been installed at third points along the length of the girders. Shear connectors had been welded on the flange of each girder in order to achieve the composite behavior with the reinforced concrete deck. The thickness of the slab was 5.1 cm, the haunch was 1.0 cm and height of the steel girders was 21.3 cm. The bridge and the rest of the structural components were designed according to the AASHTO LRFD Bridge Design Specifications [37] assuming that the bridge was located in a Seismic Zone 3. The bridge and all the connecting elements were designed and constructed at the University of Nevada, Reno and then shipped to Oregon State University.

As shown in Figure 4, the bridge was installed on steel bent caps using steel bearings, which constrained all the degrees of freedom. The bent caps were supported by a testing frame consisting of two beams and two brackets, both of which were bolted to the flume walls. The experimental setup also consisted of rails with small friction bolted on top of the black beams, carriages connected to the rails, load cells below the bent caps (shown with yellow in Figure 4) and load cells on top of the bent caps that were connected to the steel bearings below the girders (shown with green in Figure 4). This setup allowed the transfer of the loading directly to the walls of the flume while meeting the main objectives of the research project, which included the investigation of the tsunami loading applied on complex bridge decks and the distribution of the loading to the structural components (deck, cross-frames, bearings, columns) and connections. The experimental setup with the bent caps, support beams and bracket plates was initially designed by [35] to study hurricane wave loading on prestressed concrete bridges but had to be modified in this study in order to (a) be able to withstand the tsunami waves, (b) accommodate the several bridge types tested in this study, (c) replicate a bridge supported on a three column bent, and (d) allow the measurement of not only the total loads but also the individual loads in the bearings of each girder.

Figure 3. Test specimen and components during the pre-test assembly in the Large-Scale Structures Laboratory at the University of Nevada, Reno (UNR) and during the hydrodynamic testing phase in the LWF at Oregon State University.

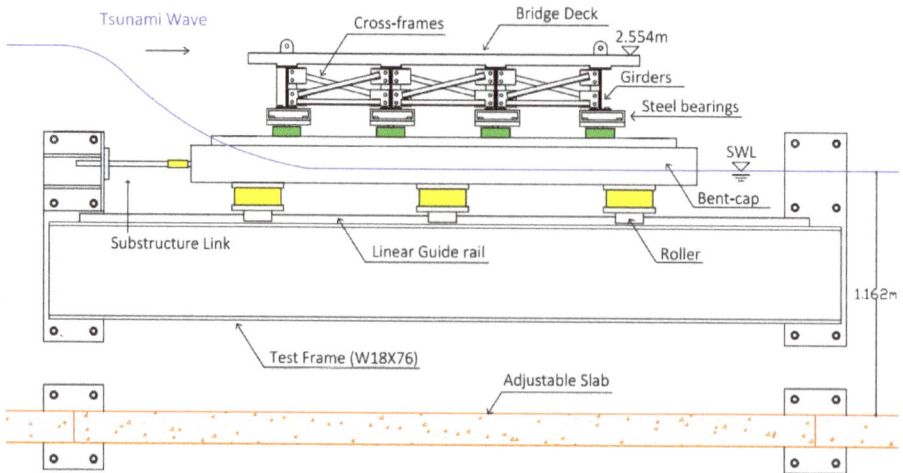

Figure 4. Cross-section of the experimental setup at the bridge location depicting the main structural components and load cells of the test case with steel bearings and rigid link.

2.2. Testing Program and Instrumentation

In the current experimental study 15 different configurations of a straight bridge (with different types of bearings and types of decks) and four configurations of a skew bridge were tested, reaching a total of 420 wave runs. This paper will discuss the experimental results obtained from the bridge with cross-frames and steel bearings with particular focus on the tsunami-induced connection forces. Several experimental studies have been conducted to date to evaluate the tsunami forces on bridges, however many of them modeled the tsunami waves as solitary waves, while other studies used broken waves or bores. The solitary waves are easier to study due to their closed-form mathematical description and repeatable wave-shape, while the broken waves/bores are more complex but at the same time can be more realistic. Therefore, in this study both types of waves were tested in order to examine the sensitivity of the bridge response to the different wave types. In both types of waves, shoaling occurs when the solitary wave propagates pass the 1:12 slope between stations 2 and 3, hence has undergone wave transformation when it reaches the location of the bridge deck. In the former the wave remains unbroken, and in the latter the wave has broken a few bays before reaching the bridge location due to the slope and the decrease in water depth. In both cases the height of the transformed wave that actually impacts the bridge is different than the nominal (or targeted) height generated at the wave-maker location. For this reason in this paper, two different wave heights will be presented, namely the nominal wave height ("H" or "Hinput") and the wave height at wave gage 12 ("Hwg12"), the closest one to the bridge, which is located at a distance of 4.44 m from the offshore face of the deck. Figure 5 shows Hwg12 for six selected heights. It should be clarified that the nominal wave height is the targeted height, however the measured heights at wg 1 relative to the targeted ones, had a difference of about 1% for the unbroken solitary waves and 1–5% for the broken ones. Table 1 shows the wave matrix, which included two water depths and a range of wave heights from 0.36 m to 1.40 m. Note that in subsequent discussions regarding solitary waves, transformation due to slope and subsequent shallower water depth is implied. The word "transformed" will be inserted when the effect of the physics of wave transformed needs to be emphasized.

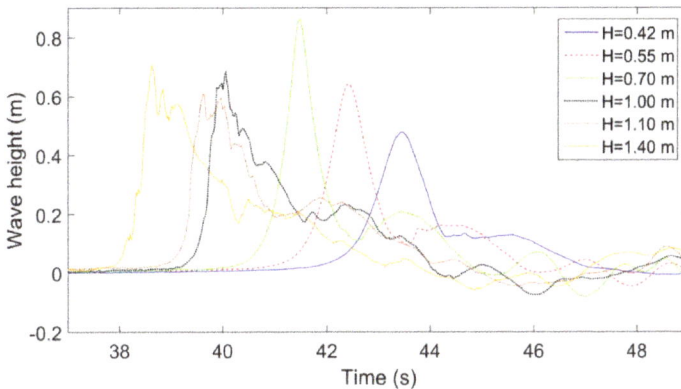

Figure 5. Wave heights recorded at wg12 for three selected unbroken solitary waves and three bores.

Table 1. Wave matrix used in the experimental testing of the bridge with steel bearings.

Water Depth (m)	Nominal Wave Height (m)	Wave Type
1.90	0.46, 0.52, 0.65 0.80, 1.00, 1.10, 1.30	Unbroken solitary Bore
2.00	0.36, 0.42, 0.52, 0.70 0.90, 1.00, 1.20, 1.40	Unbroken solitary Bore

Since the main objectives of the experimental study was the development of a high-quality database that could be used for validation and benchmarking of sophisticated numerical codes and simplified analytical methodologies, the use of extensive instrumentation was of major significance. For this reason, the wave hydrodynamics (including wave propagation, shoaling, and breaking) were measured and characterized using 13 resistive-type wave gages, 5 acoustic probes, 12 pressure gages and 16 Acoustic Doppler Velocimeters (ADVs). The wave gages were installed along the length of the flume to measure the free-surface elevation and capture the propagation of the waves (Figure 2), while the ultrasonic gages were installed at the location of the bridge to track the overtopping process. The ADVs were installed at four different locations in order to measure the flow velocities and determine the velocity profile.

The bridge was also extensively instrumented with the aim of measuring both the impact tsunami pressures and the bridge response. In particular, 12 pressure gages were installed on the steel girders and also on the concrete deck to capture the impact pressures at selected locations, while 3 biaxial accelerometers were installed at three locations on the top surface of the bridge deck. Furthermore, eight submersible load cells were installed below the girders and six submersible load cells were installed below the bent cap in order to measure the vertical forces in the girder and column-bent cap connections respectively. Furthermore, two submersible load cells were installed horizontally at the level of the bent caps to measure the total horizontal force transferred from the deck to the bent caps and the supports. Apart from the above instruments, 24 strain gages were installed on the steel cross-frames in an attempt to get an estimation of the forces carried by each member. Detailed information about the exact location of the instrumentation can be found in [23].

3. Experimental Findings

3.1. Total Horizontal and Vertical Forces

In some studies it was suggested to assume that the maximum values of horizontal and uplift forces occurred at the same time, since this was believed to be a conservative assumption for the design of the bridge [15]. Given the fact that the tsunami inundation of a bridge is a transient process, it was deemed critical in this study to investigate the transient tsunami overtopping mechanism and understand how the tsunami-induced forces change as the inundation of the bridge progresses.

For this reason, the time histories of the total horizontal forces were calculated using the measured forces in the two horizontal links and the total vertical forces were calculated by adding the vertical forces recorded in each connection. These time-histories are plotted in Figure 6 for three unbroken solitary waves and three bores of different heights. It must be noted that the origin (t = 0) of all the time series presented in the manuscript is the moment that the wave-maker starts moving, which explains why waves of different heights (and celerity) arrive and impact the bridge at a different instant. Examination of this figure reveals that for most waves the horizontal force exhibits four major peaks, which is equal to the number of the girders under the bridge deck. This indicates that the number of girders has an effect on the horizontal force histories. Regarding the vertical force histories, all tested waves introduced significant uplift forces (positive values in the graphs) as the wave hits the deck, followed by a significant and longer duration downward force (negative values). For most of the tested wave heights the deck witnessed a distinct short-duration impulsive uplift force (also called "slamming force" in other research studies) at the time of the initial impact of the wave on the deck, followed by a longer duration (slowly varying) uplift force as the chambers of the bridge started getting flooded.

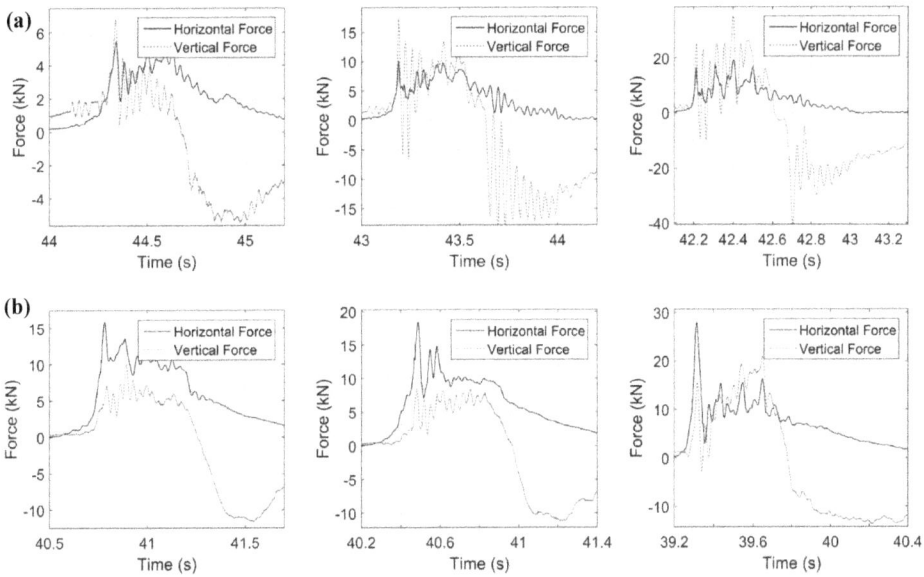

Figure 6. Total horizontal and vertical forces for (**a**) solitary waves (top) with H = 0.42 m (left), H = 0.55 m (center) and H = 0.70 m (right), and (**b**) bores (bottom) with H = 1.0 m (left), H = 1.10 m (center) and H = 1.40 m (right).

With respect to the horizontal force, the patterns were totally different for bores and unbroken solitary waves, with the bores always introducing a significant short-duration impulsive force followed by a significantly reduced in magnitude long duration lateral force. This behavior was not observed for solitary waves, which introduced several longer duration peaks with significant magnitudes during the whole inundation process. Another main difference between the two types of waves is the fact that for bores the total horizontal force was always maximized at the time of the initial impact, while for solitary waves the largest lateral force occurred either when all chambers were flooded or at the initial impact, depending on the wave height. Last but not least, all bores introduced horizontal forces larger than the vertical ones, while for the solitary waves the opposite was true. All these differences indicate that the physics describing the impact and overtopping mechanism as well as the tsunami loading on the deck are different for the two wave types, suggesting the need for the development of methodologies that will predict the tsunami-induced effects as a function of the wave type. Subsequently, this means that it is highly important to be able to identify the wave type to which a bridge at a specific location will be subjected, in order to accurately estimate the effects on the structure.

Another interesting finding that emanates directly from Figure 6 is related to the assumption made in previous studies that the maximum of the horizontal (maxFh) and vertical (maxFv) force coincide in time. Clearly as shown in the figure, this was true for some wave heights but not all. In the cases that the maximum of the two forces did coincide, it happened either at the beginning of the inundation phase (e.g., H = 0.42 m and H = 1.10 m) or later on as the bridge inundation progressed (e.g., H = 0.70 m). For most bores, the two maxima occurred at different time instants with maxFh occurring at the beginning of the inundation and maxFv occurring later on as the chambers get flooded (e.g., H = 1.40 m)

The facts that (a) the horizontal and vertical forces do not always have maximum values at the same instant, and (b) even in the wave cases where the maxima do coincide, this does not take place at the same point of the inundation process (e.g., at the initial impact of the wave on the offshore girder or at a later stage of the inundation), demonstrate the transient nature of the tsunami overtopping mechanism and its complexity, which makes it challenging to estimate exactly the tsunami effects on bridge decks. Moreover, it is noteworthy that for several wave cases where the forces in the two directions did not reach their maximum value concurrently it was observed that at the instant of the maximization of the force in one direction the force in the other direction was significantly reduced relative to its maximum value. For example, for the largest bore with H = 1.40 m at the instant of maxFh the uplift force was 75% of the maximum uplift, while at the instant of maxFv the horizontal force was 60% of maxFh. This means that making the assumption of concurrent maximum forces in the two directions might not be very accurate for several wave cases, a fact that would suggest the need for developing sophisticated methodologies, which will be able to estimate the maximum force in one direction together with the corresponding force in the other direction ultimately leading to a more economical design of tsunami resilient bridges.

3.2. Vertical Forces in Bearings

In order to be able to design a tsunami-resilient bridge the engineer must know the actual demand on each structural component and connection of the bridge. Hence the vertical forces recorded in the load cells directly below the bearings of each girder are depicted in Figure 7, together with the total vertical force for a range of (transformed) solitary waves and bores. It must be noted that in order to calculate the vertical force in the bearings of each girder (e.g., Fv, brgs, G1) the measured forces in the respective load cells of the two bent caps are added together. The previous figure yields several interesting findings and particularly:

- For both solitary waves and bores all bearings are experiencing both uplift and downward forces during the tsunami impact and the flooding process.
- At the initial impact the forces in the offshore (upstream) and onshore (downstream) bearings are out-of-phase, with the offshore bearings witnessing uplift forces and the onshore ones downward forces. At that point in time the offshore steel bearings are taking most of the uplift tsunami force with the bearings of the second offshore girder sharing part of this force. However, as the inundation progresses and the chambers become flooded then all bearing witness uplift forces that are in phase, and are sharing the total uplift force.
- The offshore bearings have to withstand significantly larger uplift forces than the rest of the bearings for all the tested waves, indicating that the offshore bearings should be designed for larger tsunami demand than the rest of the bearings in order to avoid failure of the bridge.
- The uplift forces in the bearings of the four girders are maximized at different instants (as shown by the red stars in Figure 7), highlighting the transient nature of the inundation process and the influence of the bearing forces by local effects created during the passage of the wave.
- For many wave heights and both wave types the maximum uplift force in the offshore bearings (Fv, brgs, G1) and the maximum total tsunami uplift (maxFv) do not coincide. The former forces are always maximized at the instant of the initial impact, while the latter ones can reach maximum either at the initial impact or later on as the deck flooding progresses. This is a major finding because it implies that the maximum total uplift on the deck might not result in the "worst case" scenario (largest demand) for every bearing.

The above time-histories gave an insight into the tsunami uplift demand on bearings for a group of selected wave heights. To get a better view of the recorded behavior Figure 8 presents the maximum uplift recorded in the offshore bearings versus the uplift in the same bearings at the instant where the total uplift force is maximized, for all the tested wave heights. Interestingly, this graph verifies that for many wave heights the maximum uplift force in the offshore bearings does not indeed occur when the total uplift is maximized. In particular, for 5 out of the 7 different (transformed) solitary wave heights and 5 out of 8 different bore heights tested, the uplift force in the offshore bearings at the instant of max total uplift load is smaller than the maximum uplift in the same bearings, and in fact for solitary waves this force can be down to 56% of its maximum value, while for bores the respective percentage is 51%. This suggests that the current widespread approach of using solely the total induced tsunami force as a parameter for evaluating the tsunami effects on bridges is not sufficient, since the worst case scenario (= maximum uplift force in the offshore bearings) cannot be correlated with the maximum total uplift for more than half of the tested waves.

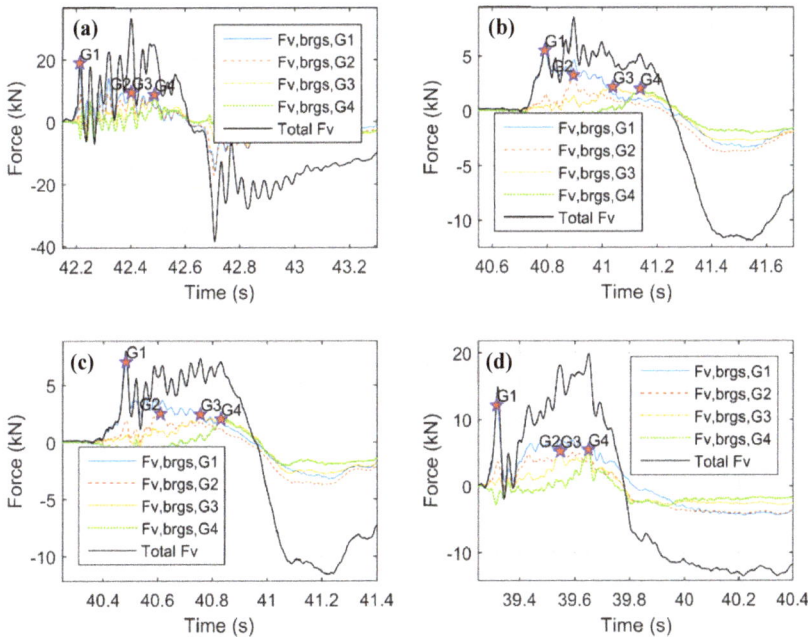

Figure 7. Vertical force histories in the steel bearings for solitary waves with H = 0.70 m (**a**), and bores with H = 1.0 m (**b**), H = 1.10 m (**c**) and H = 1.40 m (**d**).

To improve the understanding of the tsunami demand on bearings furthermore Figure 8 (bottom) shows the maximum uplift forces measured in the offshore bearings (Fup, G1) versus the maximum total uplift (Fup) for all the wave heights (Hinput) tested in this study, while Table 2 shows the ratios of these forces for all the bearings of the bridge. This graph clearly supports the previous figures that showed the offshore bearings getting larger uplift forces than the rest of the bearings. In fact, the graph reveals that the offshore bearings get a large percentage of the total uplift, which can reach 91% of the total uplift force for unbroken (transformed) solitary waves and 96% for bores, with the average values being 78% and 70% for the two wave types respectively. The forces in the rest of the bearings reach up to 41% and 37% (Fup, G2), 28% and 30% (Fup, G3), 34% and 36% (Fup, G4) of the total maximum uplift (Fup) for solitary waves and bores respectively. It is worth noting that if the tsunami uplift load was assumed to be a pure hydrostatic load and the deck was rigid then the uplift load would be equally distributed to all bearings and the bearings of each girder would have to withstand 25% of the total uplift. However, the experimental results are demonstrating that this is undoubtedly not the case here, with the offshore bearings witnessing uplift forces that can be equal to 96% of the total bridge uplift. An important issue emerging from these findings, is that since the offshore bearings are witnessing the smallest gravity load among all bearings (due to the smallest tributary areas) and at the same time the largest uplift forces, they have by far the largest probability of failure in the case of a tsunami event. To avoid such a failure, a practical recommendation for engineers designing bridges with cross-frames and steel bearings is to design the offshore bearings and connections to withstand the total tsunami uplift force.

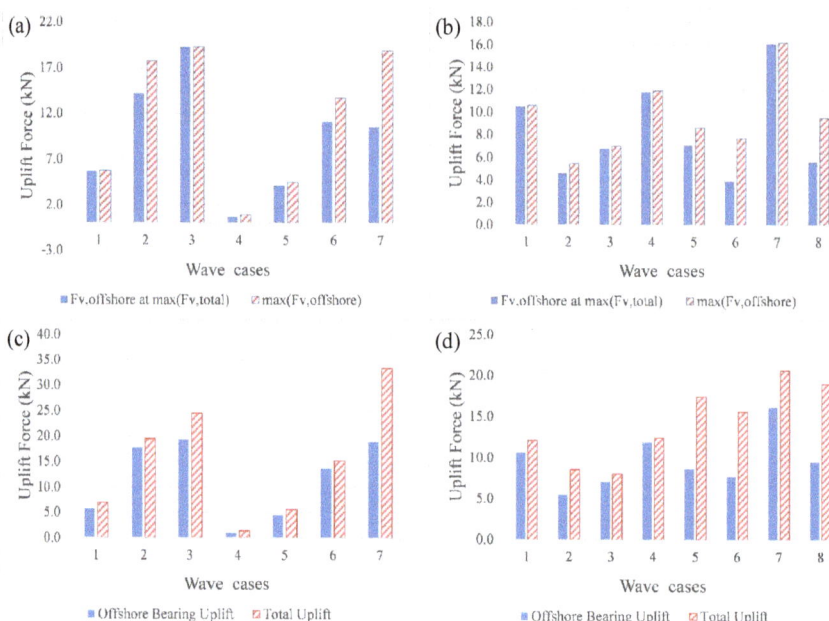

Figure 8. Uplift forces measured in the offshore bearings at time instant of maximum total uplift versus the maximum uplift forces in offshore bearings for all solitary waves (**a**) and bores (**b**), and maximum uplift forces measured in the offshore bearings versus the maximum total uplift for solitary waves (**c**) and bores (**d**).

Table 2. Ratio of experimentally recorded maximum uplift forces in each bearing relative to the maximum total uplift.

Wave Type	Depth (m)	Wave Case #	Hinput (m)	Hwg12 (m)	Fup, G1/Fup, tot	Fup, G2/Fup, tot	Fup, G3/Fup, tot	Fup, G4/Fup, tot
Un-broken solitary	1.90	1	0.46	0.56	0.72	0.23	0.12	0.09
		2	0.52	0.64	0.77	0.35	0.19	0.29
		3	0.65	0.87	0.71	0.26	0.19	0.17
	2.00	4	0.36	0.40	0.44	0.19	0.18	0.21
		5	0.42	0.48	0.67	0.21	0.14	0.09
		6	0.55	0.64	0.79	0.32	0.22	0.27
		7	0.70	0.86	0.54	0.27	0.27	0.25
Bore	1.90	1	0.80	0.85	0.86	0.23	0.20	0.21
		2	1.00	0.68	0.55	0.32	0.21	0.19
		3	1.10	0.61	0.83	0.30	0.29	0.24
		4	1.30	0.72	0.97	0.25	0.25	0.16
	2.00	5	0.90	0.84	0.47	0.26	0.27	0.31
		6	1.00	0.76	0.47	0.29	0.25	0.35
		7	1.20	0.73	0.77	0.26	0.22	0.21
		8	1.40	0.75	0.48	0.26	0.25	0.27

3.3. Vertical Forces in Column-Bent Cap Connections

If the bearings are designed to withstand the individual tsunami demand, then the failure will not occur at the interface of the superstructure and substructure of bridges as happened in past tsunamis and the tsunami loading will be transferred to the structural components below the bearings until it reaches the ground. Therefore, it is of great importance to examine and understand the demand on columns and column-to-bent cap connections. To this end, this section will present and analyze experimental data recorded in the load cells below the bent-cap (shown with green color in Figure 4), which represent the uplift forces that will have to be transferred by the column-bent cap connections. It must be clarified that the axial stiffness of the column-bent cap connections in the experimental setup is generated by the stiffness of the load cells, connecting steel plates, carriages, rails and support beam, and might not be equal to the actual axial stiffness of the columns of a bridge bent cap. This means that the actual distribution of the tsunami uplift load in the columns and column-bent cap connections might slightly differ, nonetheless it is expected that the experiments are going to provide at least a preliminary insight into the tsunami demand on the connections of a three-column bent.

Figure 9 presents the histories of the uplift forces recorded in the three connections of the bent cap together with the total tsunami uplift force. This figure presents some similarities with trends and patterns observed in the uplift forces of the bearings. In particular, the graph illustrates that:

- The offshore column-bent cap connections experience significantly larger uplift than the rest of the connections, while the onshore connections experience larger uplift than the center ones.
- The maximum uplift in the different connections/columns does not occur at the same time, and does not necessarily coincide with the maximum deck uplift. The maximum uplift demand in the offshore connections takes place at the initial impact when the onshore connections witness downward tsunami loading, while the uplift in the rest of the connections is maximized at a later instant of the inundation process.
- The offshore connections/columns have to withstand a large percentage of the total uplift force and for some waves this connection forces is larger than the total applied uplift.

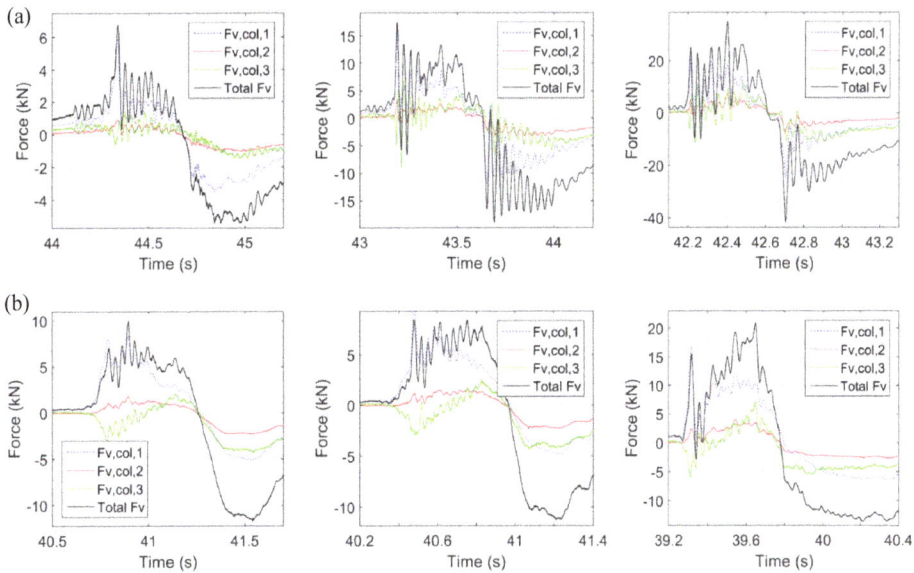

Figure 9. Vertical force histories in column-bent cap connections for (**a**) solitary waves (top) with H = 0.42 m (left), H = 0.55 m (center) and H = 0.70 m (right), and (**b**) bores (bottom) with H = 1.0 m (left), H = 1.10 m (center) and H = 1.40 m (right).

The last observation is also further reinforced by Figure 10, which shows the maximum recorded uplift in the offshore column-bent cap connections relative to the maximum of the total applied uplift for all the tested wave heights. From the graph it is verified that the uplift force in the offshore column is very large relative to the total uplift for all wave heights and both wave types, and in fact for the examined solitary waves the max uplift in the offshore column was 82% on average, while for bores the respective value was 94%. More interestingly, for 4 out of the 8 bore heights the uplift force in the offshore column-bent cap connection exceeded the total uplift tsunami force applied on the deck, by up to 24%. This behavior was observed only for bores and not for solitary waves. A possible reason behind the observed behavior is the fact that as seen in Section 3 of this paper, bores are characterized by significant lateral forces (that exceed the vertical ones), which could potentially create a large OTM as the moment arm increases. Another unexpected observation was the fact that the onshore columns experienced larger uplift than the center one, however further investigation is required in order to decipher the observed phenomenon.

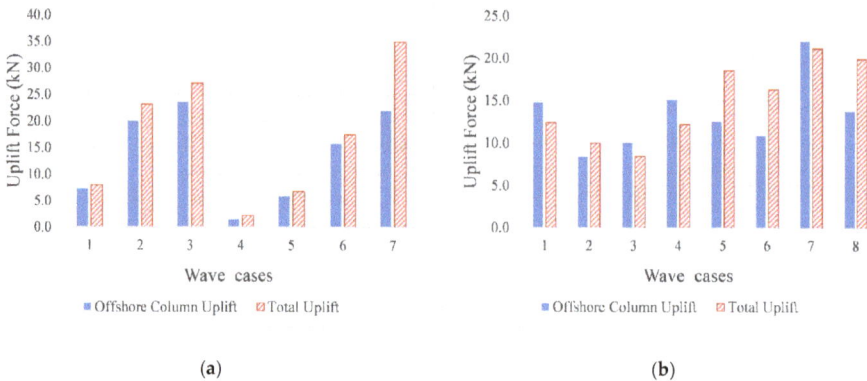

(a)

(b)

Figure 10. Maximum uplift forces measured in the offshore column-bent cap connections versus the maximum total uplift for solitary waves (**a**) and bores (**b**).

These findings suggest that contrary to most of the research studies conducted to date, which focused on the total uplift force, future studies should focus directly on the tsunami demand on connections and other structural components and consider the role of the overturning moment created by the horizontal load. Moreover, in contrast to the bearings, in which case it was suggested to design the offshore ones for the total tsunami uplift load, this cannot be done for the column-bent cap connections and columns where the overturning moment—generated by the lateral load—has a major effect. It must be clarified that the suggestion of designing the bearings for the total uplift tsunami load is expected to be applicable only to bridges with cross-frames, particularly those ones that have dimensions (e.g., girder height) and dimension ratios (height of bridge/width of bridge and width of overhang/width of bridge) similar to the bridge model tested in this experiment. For example, for a bridge with girders twice as high as the ones tested, the moment arm and consequently the overturning moment generated by the lateral load might become significant and increase the uplift forces even in the bearings to values larger than the total deck uplift.

The analysis of the experimental data highlights the importance of identifying the exact lateral load that corresponds to the maximum uplift on the deck in order to accurately estimate the overturning moment at the level of different structural components below the bent cap (e.g., column-bent cap connections, top and bottom of columns, column-foundation connection, etc.) and consequently the tsunami uplift demand on specific bridge members. The fact that the offshore columns and column-bent cap connections experience significant uplift force with simultaneously large horizontal force and possibly large concentrated moments, coupled with the fact that reinforced concrete sections have reduced moment and shear capacity for large tensile forces, imply that the offshore column-bent cap connections or actual columns might fail in a tsunami event if the bearing connections are strong enough to transfer the applied load from the deck to the bent cap.

4. Estimation of Bearing and Column Uplift Forces Based on Simplified Approach

4.1. Application of Maximum Horizontal and Vertical Forces

Azadbakht and Yim [15] investigated the tsunami impact on bridge decks via 2D numerical analyses and utilized the results for the development of simplified equations that estimated the maximum total lateral, uplift and downward forces. Their research study argued that although the maximum horizontal load and maximum uplift load applied on the bridge deck might not occur at the same time, it would be conservative to apply them simultaneously at the center of gravity (CG) of the deck. Other research studies [21] assumed that the hydrodynamic component of the tsunami loading is uniformly applied on the face of the bridge, meaning that in this case the resultant horizontal load would be applied at the mid-height of the bridge. This approach seems to be consistent with FEMA P-646 [38], which suggests that the hydrodynamic loading applied on the face of a building is uniform with height and the resultant force is applied at the mid-height. Other assumptions could be made regarding the location of application of the lateral load such as the mid-height of the girder or the CG of the concrete slab. Calculating the CG of the concrete slab is generally simpler than the CG of the whole deck, so applying the force in the former location might be preferable, however the final location of application should be based on the physics involved rather than simplicity.

Once the applied load is known and the location of application is selected then a free body diagram (Figure 11) can be drawn to assist with the estimation of the reaction forces. Figure 11 assumes a generic load case, where the bridge is subjected to horizontal and vertical forces as well as an overturning moment. Given the number of unknowns, the system is indeterminate and calculation of the reaction forces would require a 3D numerical model of the bridge deck that would simulate the actual stiffness of every structural component and capture the force distribution accurately. Alternatively, for the sake of simplicity and with the aim to get a first order estimation of the reaction forces the bridge can be assumed rigid and the reaction moments equal to zero. In such case it is possible to solve the static equilibrium and estimate the uplift forces in the bearings using the following equations:

$$Fv1 = (Fv/4) + (3/10) \times (Lz/Lx) \times Fh + (3/10) \times (M/Lx), \tag{1}$$

$$Fv2 = (Fv/4) + (1/10) \times (Lz/Lx) \times Fh + (1/10) \times (M/Lx), \tag{2}$$

$$Fv3 = (Fv/4) - (1/10) \times (Lz/Lx) \times Fh - (1/10) \times (M/Lx), \tag{3}$$

$$Fv4 = (Fv/4) - (3/10) \times (Lz/Lx) \times Fh - (3/10) \times (M/Lx), \tag{4}$$

In these equations Fh is the total tsunami horizontal load, Fv is the total applied uplift load, M is the applied moment (assumed zero), Lx is the distance between the bearings of two consecutive girders and Lz is the distance from the location of application of Fh and the center of the load cells below the bearings. Given the fact that past research studies applied the horizontal load at different locations vertically, in this study three different scenarios were examined with Fh being applied at (a) mid-height of the bridge (Lz = 0.241 m), (b) CG of bridge deck (Lz = 0.313 m), and (c) CG of concrete slab (Lz = 0.347 m). Notably, the larger the Lz, the larger the calculated uplift forces in the offshore bearings and bearings G2. Similar equations with Equations (1)–(4), can be developed for calculating the reaction forces in the three columns and connections of the bent cap; however, due to space limitations these equations are not shown herein. For the purpose of possible reproducibility it is noted that for the calculation of the uplift forces in the column-bent cap connections, Lx = 0.914 m and Lz = 0.552 m, 0.625 m, 0.659 m for the three aforementioned scenarios.

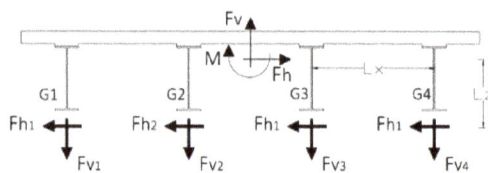

Figure 11. Free body diagram showing applied loads and reaction forces for the simplified method.

Using the simplified methodology together with the experimentally measured total lateral and uplift load, the uplift forces in the bearings and column-bent cap connections have been calculated. These calculated component uplift forces, are then compared to the experimentally measured forces in the same components, in order to assess the accuracy of the simplified methodology. Table 3 presents the ratios of the calculated (using the simplified method) to the experimentally measured uplift forces in all bearings and column connections for all the tested wave heights assuming that the lateral load is applied at the CG of the deck. A positive ratio smaller than 1 implies that the method under-predicts the uplift force, a ratio larger than 1 implies over-prediction and a negative ratio indicates that the simplified method fails to predict the right direction of the loading (downward load instead of uplift). It becomes apparent from the table that the simplified method under-predicts the uplift forces in the offshore bearings (G1) for all wave heights, by up to 59% and 30% on average. For bearings G2, the method gives better results when it over-predicts the uplift forces by 16% on average and 59% max, and under-predicts the force for only three wave heights. For bearings G3 and G4, the simplified approach again fails to predict the measured uplift forces and it actually under-predicts the demand by up to 67% and 98% for the two bearings respectively, while in some cases for the onshore bearings it fails to even predict the correct direction. This may be due to the assumptions made in this methodology such as the rigidity of the bridge deck or the application of maxFh and maxFv at the CG of the deck. It must be noted that the latter assumption, which was recommended in reference [15] was based on tsunami waves with horizontal prescribed velocity at the inlet and wave propagation over a flat bathymetry, without accounting for shoaling effects. Due to these effects the solitary wave transforms and increases in height before it reaches the bridge, a transformation that could generate a vertical velocity component. This in turn could influence the induced forces and their location.

Table 3. Ratio of bearing and column-bent cap connection uplift forces calculated from the simplified method to the respective values recorded in the experiments.

Wave Type	Hinput (m)	Bearings, G1	Bearings, G2	Bearings, G3	Bearings, G4	Col. 1	Col. 2	Col. 3
Un-broken solitary	0.46	0.52	1.17	1.30	0.63	0.70	2.12	0.19
	0.52	0.41	0.71	0.92	0.37	0.61	1.58	0.34
	0.65	0.54	1.06	0.91	0.40	0.73	1.87	0.11
	0.36	0.85	1.25	0.49	−0.24	1.08	1.90	−0.17
	0.42	0.53	1.23	1.15	0.62	0.70	2.43	0.27
	0.55	0.41	0.79	0.84	0.42	0.59	1.75	0.33
	0.70	0.63	1.02	0.77	0.56	0.83	1.55	0.45
Bore	0.80	0.73	1.59	0.59	−0.67	0.89	2.22	−1.91
	1.00	0.92	0.97	0.56	−0.39	1.05	1.87	−1.05
	1.10	0.77	1.25	0.37	−0.68	0.91	1.69	−1.37
	1.30	0.67	1.53	0.49	−0.87	0.88	2.03	−1.94
	0.90	0.95	1.18	0.62	0.09	1.07	1.64	−0.17
	1.00	0.99	1.11	0.65	0.02	1.15	1.83	−0.28
	1.20	0.68	1.29	0.70	−0.14	0.82	1.82	−0.71
	1.40	0.95	1.21	0.67	0.09	1.07	1.77	−0.21

Similar trends are observed for the uplift forces in the column-bent cap connections, with the simplified method giving good estimates for the center column (col. 2), but under-predicting for most waves the demand on the offshore one (col. 1) by up to 41% (and 13% on average) and predicting a wrong direction for the vertical forces in the onshore column (col. 3) An important issue emerging from the above observations, is that although the simplified method of applying the total tsunami forces at the CG of the deck seems to be giving good results for the uplift forces in the structural components and connections close to the CG (particularly bearings G2 and center column), it fails to accurately estimate the demand on components far from the CG and especially offshore and onshore bearings and columns. In fact for these components the simplified method yields significant under-predictions of the uplift force demonstrating that it is not conservative as initially hypothesized. This finding was unexpected and suggests the need for further investigation and ideally development of an improved practical methodology that will be able to predict the tsunami uplift in all structural components conservatively.

4.2. Role of Overturning Moment

Since the application of the total horizontal and uplift tsunami load at the CG of the deck cannot estimate accurately the uplift demand on bearings and columns, it means that the physics of the tsunami impact and inundation of the bridge are not accurately represented in the simplified methodology. Going back to the time-histories of the uplift forces in the connections (Section 3), one can observe that at the beginning of the tsunami impact the vertical forces in the offshore and onshore bearings are out-of-phase, with the former experiencing uplift and the latter downward force. This implies the existence of a clockwise overturning moment, which could potentially increase the demand on offshore connections and decrease it on the onshore ones. To examine if this is actually the case, the moment histories have been calculated directly from the experimental measurements and will be presented in this section.

Figure 12 shows a free-body diagram of the deck alone as well as the deck together with the bent-cap, in order to assist the calculation of the overturning moment at the level of the bearings and the level of the column-bent cap connections. In reality the tsunami load applies transient pressures on different locations of the bridge and the applied horizontal and vertical tsunami loading changes both spatially and temporally during the inundation process. However, at each instant the applied tsunami moment shall be equal to the reaction moment in the connections plus the rotational inertia and damping force of the system in order to satisfy equilibrium. If the inertia and damping forces are assumed to be negligible, then the applied moment will be equal to the reaction one. The latter moment can be calculated using the reaction forces and moments in the bearings (or in the bent-cap connections) as shown in Equation (5).

$$\text{Moment} = -(Fh1 + Fh2 + Fh3 + Fh4) \times Lz + Fv1 \times L1 + Fv2 \times L2 - Fv3 \times L3 - Fv4 \times L4 + \Sigma Mi \quad (5)$$

In this equation $Fh1 + Fh2 + Fh3 + Fh4$ is equal to the horizontal force recorded in the links, Fvi is the vertical force recorded in bearing i, Li = horizontal distance of bearing i from the point about which the moment is calculated, Lz = vertical distance of the center of the load cells from the point about which the moment is calculated and ΣMi is equal to the sum of the reaction moments ($M1$, $M2$, $M3$, $M4$). Since the connection moments are expected to be small they could be neglected, in which case the applied moment is estimated directly from the experiment. It is noteworthy that the moment was also calculated using the measurements in the load cells below the bent cap and then compared with the moment calculated using the bearing measurements and good agreement was observed, as shown in [23] increasing the confidence in the calculated moment histories. These histories are plotted in Figure 13 together with the total horizontal and vertical forces, for selected solitary waves and bores. From the figure it becomes apparent that:

- Significant clockwise moment is generated at the instant of the first impact of the wave on the offshore face of the bridge. This moment is reduced as the wave propagates through the bridge and later on it becomes counter-clockwise.
- Although the forces can be maximized at different instants during the deck flooding, the moment is always maximized at the initial impact of the wave on the offshore girder and overhang for both wave types.

- The maximum overturning moment and maximum bridge uplift do not occur at the same time for many of the tested waves in the experiment. The existence of this significant overturning moment could possibly explain why for the solitary wave of H = 0.70 m although the maximum total uplift force occurs when the wave has flooded all the chambers, the maximum uplift in the offshore bearings occurs at the time of the initial impact.

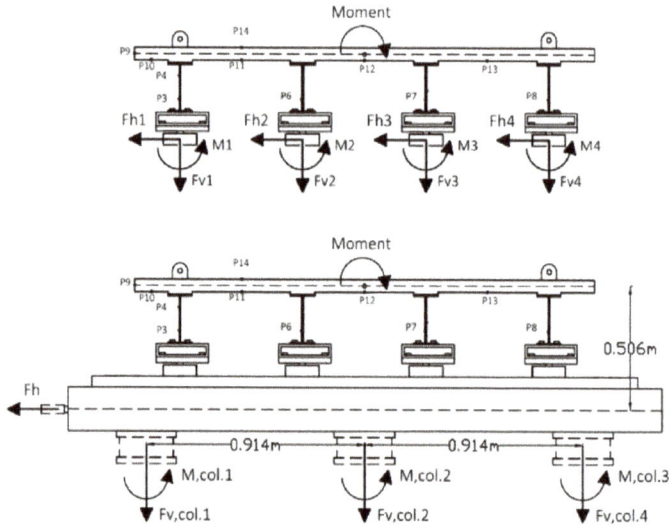

Figure 12. Calculation of tsunami-induced moment at the level of the bearings (top) and the level of the column-bent cap connections (bottom).

The time-histories revealed the existence of a significant OTM, however in order to understand its significance for the bridge bearings and connections, further investigation is needed. To that end, the maximum recorded uplift forces in each bearings are plotted in Figure 14 relative to the maximum recorded total tsunami uplift and the maximum OTM, revealing that:

- The offshore bearings do not have a good correlation with the maximum recorded total uplift (R^2 = 0.62), and in some cases a larger total uplift gives a smaller uplift in the bearings, which seems counter-intuitive. Instead, the uplift in these bearings is better correlated with the maximum clockwise moment (R^2 = 0.83). This is a major finding and it seems to make sense if one considers the observations made in previous sections, according to which the uplift force in the offshore bearings was always maximized at the first impact of the wave on the bridge, while the maximum total uplift on the deck was maximized at different instants of the inundation process depending on the wave height.
- The bearings of the internal girders (G2 and G3) are showing good correlation with the maximum uplift on the deck, with G2 having the best agreement among all bearings (R2 = 0.926). For both of these bearings the correlation with the max clockwise moment is poor, indicating that the moment does not govern tsunami-induced uplift at these locations.
- Generally, as the horizontal distance of the bearings from the offshore face of the bridge increases, the agreement of the individual bearing uplift with the maximum clockwise moment weakens, and for the onshore bearings there is no apparent correlation (R^2 = 0.33). The onshore bearings also seem to have a weaker correlation with the maximum total uplift than the bearings of the internal girders, demonstrating that predicting the maximum uplift in these bearings might be more complex than expected.

To investigate whether the overturning moment is the reason behind the large uplift in the offshore bearings, it was decided to modify the simplified method, which applied the total maxFh and maxFv at the CG of the deck, and include the maximum clockwise moment Mmax (assuming they all occur simultaneously) to see if a better prediction of the uplift forces in the bearings and connections could be achieved.

Using these equations, the maximum uplift in each bearing was calculated and then compared with the experimentally measured values. Table 4 presents the ratios of the estimated to the measured maximum uplift in all bearings and columns. As shown, the improved simplified method with the applied moment gives conservative uplift forces for bearings G2 and for the offshore and center columns. For the offshore columns the method gives an overestimation of 16% in average and 41% maximum. For the offshore bearings the method gives an overestimation of 4% on average, and a difference between −11% and +29% relative to the measured values. It is worth noting that the simplified approach of considering only maxFh and maxFv without the moment gave under-prediction of the uplift force in the offshore bearings by 30% on average and a difference between −59% and −1% relative to the measured values. This provides concrete evidence regarding the significance of the tsunami-induced overturning moment and suggests that the simplified method presented previously most likely did not accurately estimate the demand on the components because it did not properly simulate the overturning moment. Although the inclusion of the moment gives much better results for most of the bearings and columns, it still cannot predict the uplift forces on all connections and, particularly, bearings G3, G4 and the offshore column.

These findings will have significant implications for the development of tsunami resilient bridges and design guidelines because they demonstrate the necessity to develop methodologies that will be able to predict not only the maximum total applied tsunami forces, which have been the focus of most of the research studies available in the literature, but also the overturning moment in order to accurately calculate the demand on individual structural components.

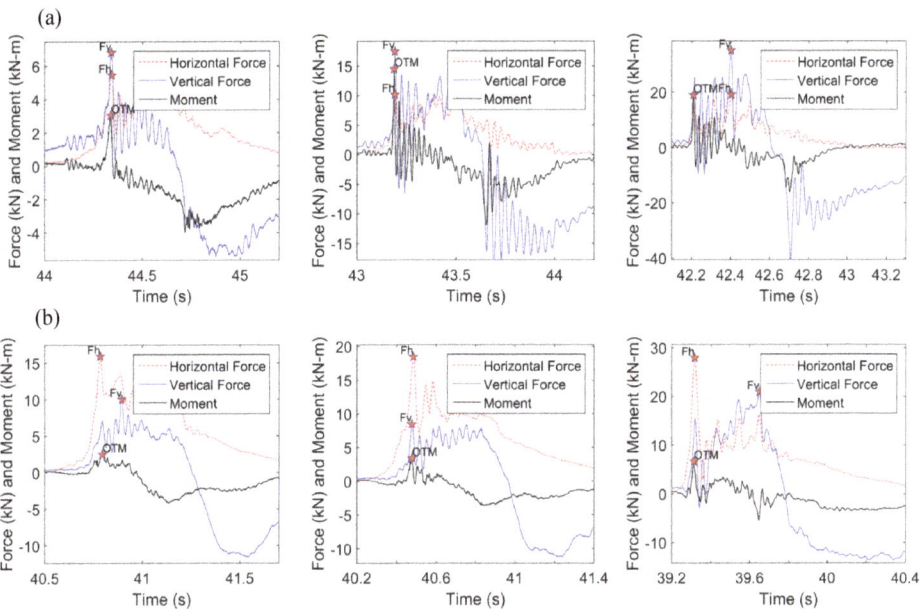

Figure 13. Total horizontal forces and moment for (**a**) solitary waves (top) with H = 0.42 m (left), H = 0.55 m (center) and H = 0.70 m (right), and (**b**) bores (bottom) with H = 1.0 m (left), H = 1.10 m (center) and H = 1.40 m (right).

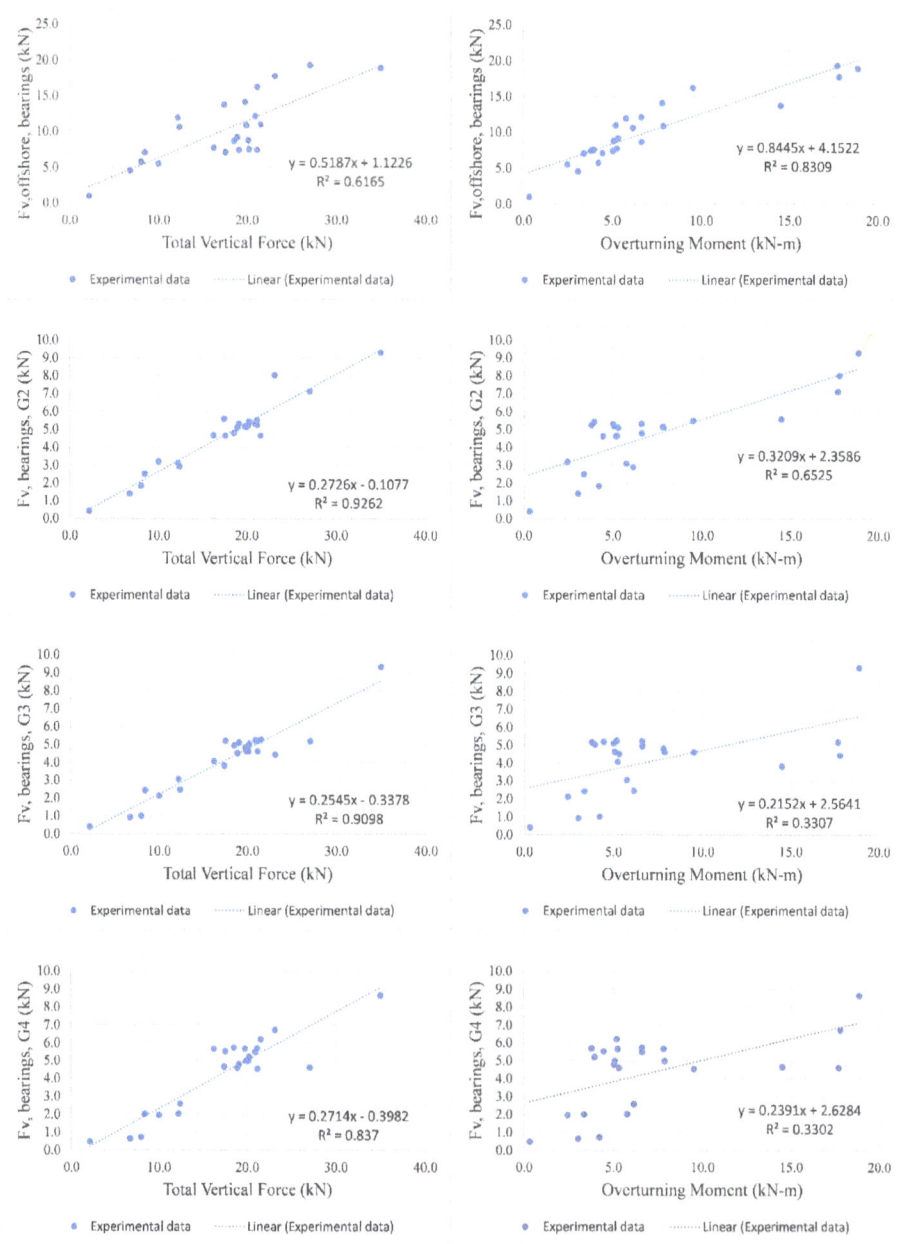

Figure 14. Measured maximum uplift in individual bearings as a function of maximum total uplift (**left**) and overturning moment (**right**).

Table 4. Ratio of bearing and column-bent cap connection uplift forces calculated from the improved simplified method (with applied moment) to the respective values recorded in the experiments.

Wave Type	Hinput (m)	Brngs, G1	Brngs, G2	Brngs, G3	Brngs, G4	Col. 1	Col. 2	Col. 3
Unbroken solitary	0.46	0.93	1.61	0.51	−2.66	1.01	2.12	−1.36
	0.52	0.91	1.07	0.26	−0.94	1.10	1.58	−0.70
	0.65	0.96	1.44	0.38	−1.37	1.14	1.87	−1.02
	0.36	1.06	1.40	0.33	−0.66	1.20	1.90	−0.36
	0.42	0.90	1.64	0.53	−2.05	0.98	2.43	−0.89
	0.55	0.89	1.18	0.26	−0.99	1.09	1.75	−0.80
	0.70	1.13	1.36	0.43	−0.53	1.30	1.55	−0.46
Bore	0.80	0.98	1.89	0.23	−1.69	1.12	2.22	−3.24
	1.00	1.16	1.11	0.35	−1.08	1.21	1.87	−1.70
	1.10	0.99	1.46	0.15	−1.46	1.09	1.69	−2.08
	1.30	0.97	1.91	0.10	−2.65	1.09	2.03	−3.13
	0.90	1.29	1.38	0.42	−0.43	1.36	1.64	−0.74
	1.00	1.25	1.25	0.48	−0.34	1.41	1.83	−0.75
	1.20	0.97	1.58	0.36	−1.19	1.05	1.82	−1.66
	1.40	1.23	1.39	0.49	−0.43	1.29	1.77	−0.65

5. Tsunami Demand Diagrams

In contrast to other extreme natural hazards such as earthquakes (EQs), where the applied loading in one direction is dictating the demand on the structural components of a bridge (e.g., for EQs the horizontal inertia forces are dictating the demand), for tsunamis this is not the case. In fact, as shown in previous sections of this paper, tsunami-like waves apply simultaneously large forces both in the horizontal (Fh) and vertical (Fv) direction as well as an overturning moment (OTM). This could possibly explain why so many bridges in the 2011 Great East Japan Earthquake were able to withstand the earthquake shaking but were damaged by the following tsunami. The time-histories of Fh, Fv and OTM gave an insight into the transient nature of the tsunami inundation mechanism and the temporal variation of the induced effects; however, due to the complexity of the phenomenon and its dependence on the wave type and wave height it is hard to decipher the physics involved and reach general conclusions. Hence, in an attempt to develop a more comprehensive understanding of the tsunami-induced effect the recorded forces and the moment were plotted against each other, developing what will be called in this paper "tsunami demand diagrams". Such diagrams are shown in Figures 15 and 16, with the former figure showing the total recorded horizontal force versus the total vertical force and the latter one showing the overturning moment versus the total vertical force. It should be stated that the concept of such a diagram is not new for structural engineers, who for decades have used interaction diagrams of axial force and moment as a means for determining the capacity of reinforced concrete columns.

Figures 15 and 16 demonstrate a complex temporal variation of Fh versus Fv as well as OTM versus Fv (blue color in the graphs), which is different for each wave height, making it very challenging (if not impossible) to predict this transient behavior via simplified approaches. Alternatively, it could be simpler and more practical to develop envelopes of Fh versus Fv and OTM versus Fv and get a sense of the maximum demand. To that end, the experimental results were further analyzed in Matlab [39] and three different envelopes with different refinement levels (refined, medium, simplified) were developed using the "boundary" function. This function was preferred over the "envelope" function available in Matlab because it allows the specification of a shrink factor s in order to adjust the compactness of the generated boundary. For example, for s = 0 the function will yield the convex hull while for s = 1 the result will be a compact boundary that envelopes the points. Therefore, the strict definition of the technical term "envelope" used in engineering, would require an s = 1. For the "Env-refined" shown in the figures an s value equal to 0.95 was used, which is close enough to 1 and computationally less expensive. For the other two boundaries "Env-medium" and "Env-simplified" the selected value for s was 0.25 and 0.05 respectively. Figures 15 and 16 demonstrate that:

- For most solitary waves, the maximum horizontal force and maximum uplift force occur at the same time, while for most bores the two maxima do not coincide (Section 3).
- All waves introduce significant clockwise moment (positive value on graph) and simultaneously large uplift loading, however for most wave heights the maxOTM does not take place at the same instant with maxFv. This implies that for the tsunami design of bridges engineers will have to consider several load cases (e.g., maxOTM with corresponding Fv and Fh or maxFv with corresponding OTM and Fh) in order to identify the worst-case scenario for each component.
- For all wave heights a large counter-clockwise moment is generated during the inundation, which was not expected beforehand. In fact, for solitary waves at the instant of the maximum counter-clockwise moment the wave applies a downward vertical load, while for bores this is not true and significant uplift is observed instead.
- The refined envelopes (red color) of Fh versus Fv and OTM versus Fv are quite complex and their shapes have significant differences between waves indicating a dependence on the wave height and wave type. However, as the envelope becomes less refined then similarities start appearing between the different heights. Especially, the simplified envelopes (Env-simplified) of Fh versus Fv and the ones of OTM versus Fv have very similar shapes among different bore heights. This tends to be true for the unbroken solitary waves too but the similarities are not as striking.

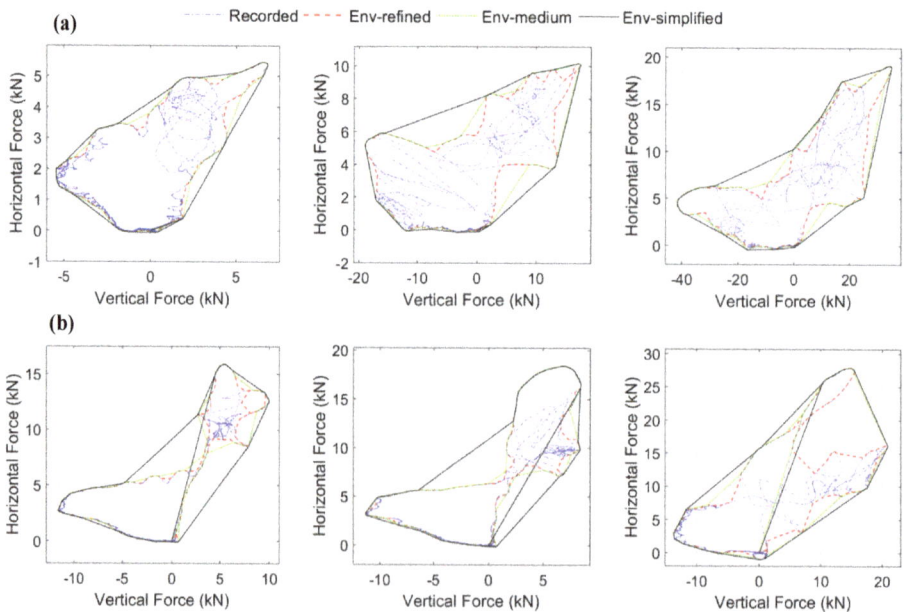

Figure 15. Demand diagrams of vertical and horizontal force for (**a**) solitary waves (top) with H = 0.42 m (left), H = 0.55 m (center), H = 0.70 m (right), and (**b**) bores (bottom) with H = 1.0 m (left), H = 1.10 m (center) and H = 1.40 m (right).

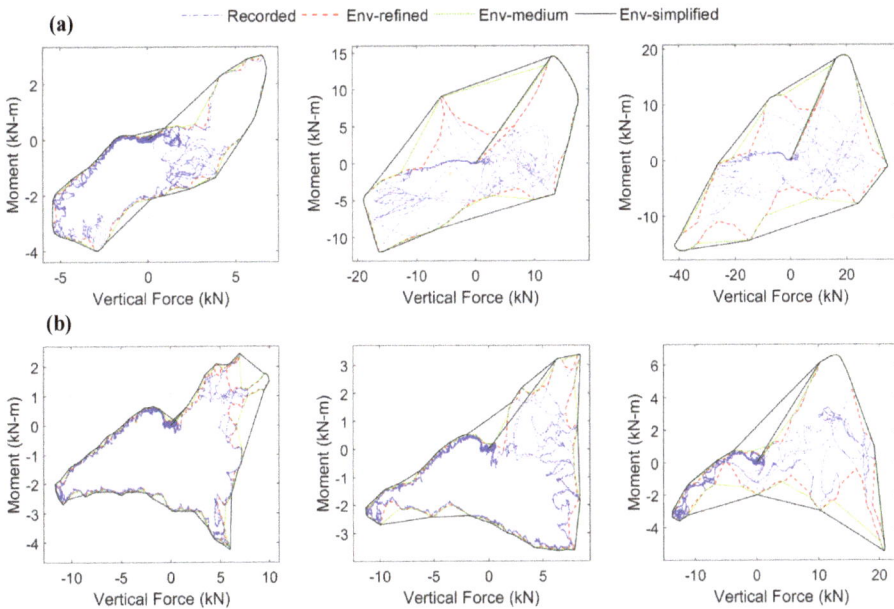

Figure 16. Demand diagrams of vertical force and moment for (**a**) solitary waves (top) with H = 0.42 m (left), H = 0.55 m (center) and H = 0.70 m (right), and (**b**) bores (bottom) with H = 1.0 m (left), H = 1.10 m (center) and H = 1.40 m (right).

It is noteworthy that these 2D demand diagrams add to the understanding of the tsunami-induced loading. However, they are expected to have a direct application in engineering practice only for cases where the design of the connections and structural components depends on two of the three parameters (Fh, Fv, OTM). For example, the 2D diagrams could potentially be used for the design of bridges with elastomeric bearings and shear keys because in such cases the shear keys would be designed to take the horizontal load and the elastomeric bearings would be designed for the uplift load using the OTM versus Fv demand diagrams. For other types of bridges though, such as the ones with steel bearings, where the bearings have to withstand both the horizontal and vertical loading, 3D demand diagrams that represent combinations of (Fh, Fv, OTM) would have to be used.

In this study three different sets of 3D demand diagrams have been developed for each wave height using the exact same values of the scalar factor s used in the 2D diagrams, and the simplified envelope is presented in Figure 17 for several wave heights. In the 3D case the "boundary" function in Matlab returns a triangulation representing a single conforming 3-D boundary around the points (Fh, Fv, OTM). As observed in the graph, all simplified 3D demand diagrams have diamond-like shapes and the ones of the bores have similarities in shape but are different in magnitude for most wave heights. This also seems to be true for the solitary waves. It could be argued that in the same way that simplified 3D analytical interaction diagrams (N, Mx, My) have been developed in the past for estimating the capacity of RC concrete sections by determining just a few points on the diagram, a similar thing could be done for estimating the tsunami-induced loading in terms of (Fh, Fv, OTM). These findings are advancing the understanding of the tsunami-induced loading on bridges and are expected to have significant implications for the tsunami design of bridges because they demonstrate:

1. The existence of fundamental differences in the effects introduced on the bridge by the two different wave types, which suggests the need for the development of methodologies that will be able to predict the exact wave type at a particular bridge location.

2. The need to consider different load cases/combinations of (Fh, Fv and OTM) and not just maxFh and maxFv as done to date, since it is not a priori known which case will be governing the design of individual structural components of the bridge.
3. The possibility to develop simplified 2D demand diagrams (Fh versus Fv, OTM versus Fv) or simplified 3D diagrams (Fh versus Fv versus OTM), which will have the same shape for all bores and a size that will change with the wave height. Once the shape is known then simplified predictive equations could potentially be developed for estimating the magnitude. This approach sounds quite futuristic for the time being; however, it is simpler than trying to predict both the spatial and temporal variation of the applied forces. Such an approach would be less economical but more convenient. Given the uncertainties involved in the tsunami wave breaking and impact on structures, some conservatism in the method that estimates the tsunami-induced loading on the structure might be acceptable.

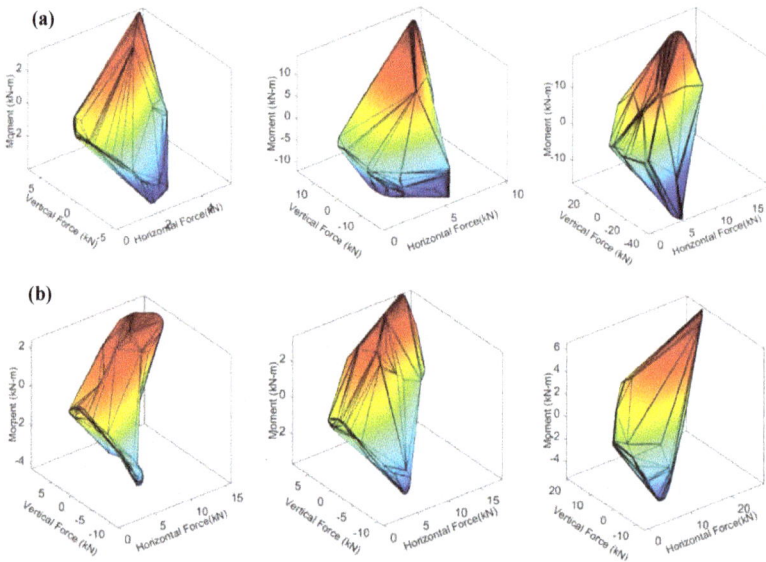

Figure 17. 3D demand diagrams of forces and moment for (**a**) solitary waves (top) with H = 0.42 m (left), H = 0.55 m (center) and H = 0.70 m (right), and (**b**) bores (bottom) with H = 1.0 m (left), H = 1.10 m (center) and H = 1.40 m (right).

6. Tsunami Inundation Mechanism of Bridges

The detailed analysis of the experimental data presented in the previous sections has demonstrated the complexity of the tsunami-induced loading. To decipher this complexity it is essential to understand the physical phenomenon that generates the loading. To this end, video recordings were processed and selected snapshots are presented in Figure 18. These snapshots show that after the wave impact on the offshore girder and overhang, part of the wave propagates below the deck flooding the chambers (as indicated by the air and water that escapes from the sides of the bridge), while the top part of the wave that hits the front face starts splashing, breaking and slamming on the top of the bridge deck until it inundates it. Interestingly, it is revealed that the wave propagates below the deck (and floods the chambers) faster than it moves above the deck, probably due the fact that the overtopping process is delayed by the splashing. This could explain why the vertical force histories exhibit first a significant uplift force followed by a downward one.

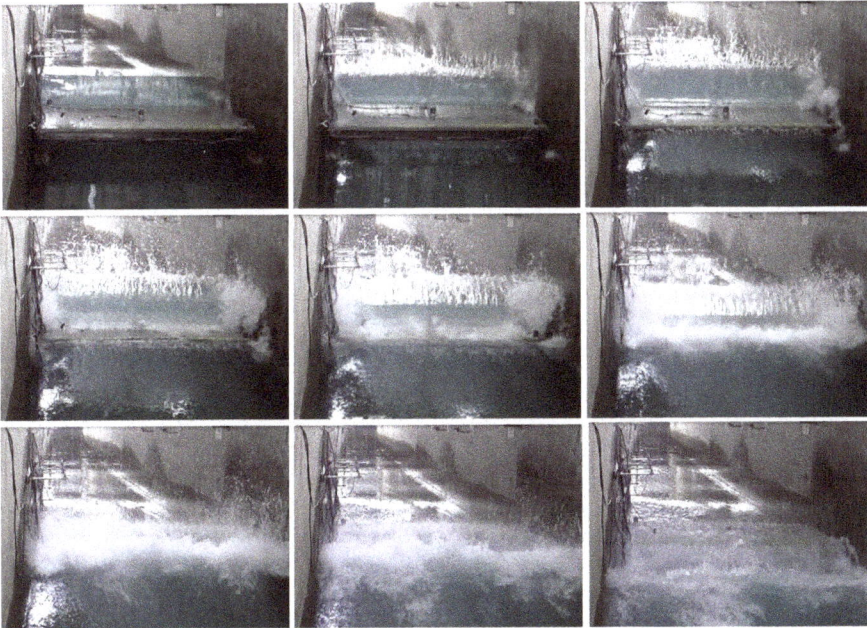

Figure 18. Snapshots during the wave impact and inundation of the bridge deck.

To decipher the bridge inundation mechanism and induced effects, it was critical to relate the experimental data from load cells, pressure gages and ultrasonic gages at different locations of the bridge. Figure 19 shows the locations of five ultrasonic gages, with two of them (uswg1 and uswg5) recording the wave height slightly before and after the bridge location (at a horizontal distance of 0.92 m and 0.82 m from the nearest bridge point respectively), and the other three gages recording the water height that overtops the deck. Moreover, the same figure shows the location of all pressures gages, with several gages recording the pressure on the girders (in the horizontal direction), below the offshore overhang of the bridge (P10), and below the deck in chambers 1, 2 and 3 (P11, P12 and P13 respectively). The recorded pressures and water height histories were normalized with their respective maximum values and are plotted in Figure 20 together with the force and overturning moment histories. In particular, the graph with the forces depicts the vertical forces in each bearing, the total induced horizontal and vertical force as well as the moment, while the graph with the normalized pressures shows the pressures below the deck at different locations starting from the offshore overhang and continuing with the chambers in order to get a sense about the spatial variation of the applied loading and location of the wave during its propagation through the bridge. Simultaneous examination of these three graphs reveals the existence of four main phases in the bridge inundation mechanism, as described below:

- **Phase 1:** This is a short duration phase during which the bridge experiences significant impulsive horizontal and vertical forces and occurs when the wave hits the offshore girder and overhang. This phase produced the maximum horizontal force for all bores and the maximum overturning moment, generating an uplift and a downward force in the offshore and onshore bearings respectively. Due to the simultaneously large uplift force and moment, this phase produces the largest uplift in the offshore bearings meaning that it could be the most catastrophic phase for the offshore structural components (bearings, connections, columns etc.).
- **Phase 2:** This is a longer duration phase that starts when the wave reaches chamber 1 and finishes when the wave has inundated chamber 2 of the bridge, at which point the uplift force is applied

close to the CG of the deck. During this phase the overtopping process begins, with the tip of the wave splashing over the top of the offshore overhang. In this phase, the applied moment is small, however the bridge uplift force is very large and for some wave heights it can actually reach its maximum value. In this phase all bearings are in phase witnessing uplift forces, and the bearings of the interior girders (G2 and G3) are reaching their maximum uplift values for most waves. This phase can also produce the largest horizontal force for solitary waves, however this is not true for bores, the horizontal force of which is reduced significantly since the slamming component is minimized and only the hydrodynamic (drag) component remains.

- **Phase 3:** This phase occurs when the wave has reached chamber 3. During this phase the tip of the wave starts inundating the offshore side (uswg2) of the top surface of the deck and towards the end of the phase the water reaches the onshore side (uswg4). In this phase the horizontal force is reduced for all wave types, the vertical force is large -can even be at a maximum- and there is a counter-clockwise moment, which generates the largest uplift force in the onshore bearings. Therefore, this phase has to be considered because it is the governing one for the onshore bearings. It must be noted though that this phase is different for bores and solitary waves because the former waves introduce the maximum counter-clockwise moment in this phase, while this is not true for the latter waves.

- **Phase 4:** This phase occurs after the wave has passed the onshore chamber, the whole top surface of the deck becomes inundated and the overtopping water introduces large downward pressures on the deck (at P14). The slamming downward force on the deck is so high that it exceeds the simultaneously applied uplift force below the deck, consequently introducing compression in all bearings. This phase also produces the maximum counter-clockwise moment for all solitary waves.

The four phases are illustrated in Figure 21, which shows a schematic of the applied pressures, the direction of the vertical forces in the bearings (brown and green arrows), and the direction of the total forces and the overturning moment. Special attention should be given to Phase 1, which has been neglected to date because it introduced the largest tension in the offshore bearings for most waves. This is of utmost significance because it indicates that if the offshore bearings exceed their tensile capacity in Phase 1 due to the significant overturning moment and uplift load and get damaged, then the uplift loads would be redistributed to the remaining bearings with the possibility of leading to a progressive collapse mechanism that will eventually result in the washout of the bridge. Moreover, Phase 3, which has also not been investigated in previous research studies should also be considered in order to accurately estimate the uplift demand on the onshore bearings, columns and connections. The characteristic counter-clockwise moment generated in Phase 3 can also explain the observations made in previous sections according to which the onshore column-bent cap connections were seen to be getting larger uplift forces than the center ones for many of the tested wave heights.

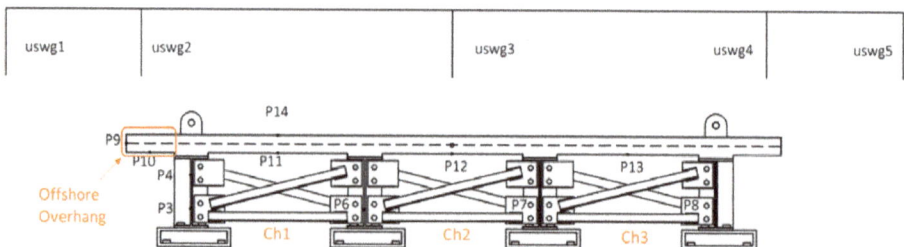

Figure 19. Schematic of bridge deck with location of pressure gages and ultrasonic wave gages.

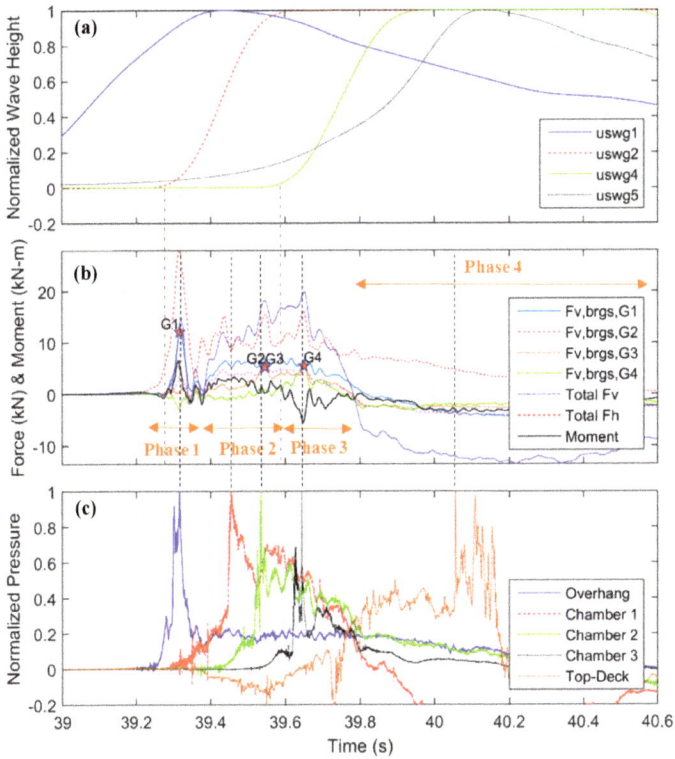

Figure 20. Tsunami-induced bearing forces and moment histories (**a**), normalized pressure histories on and below the deck (**b**), and normalized wave height (**c**), for H = 1.40 m.

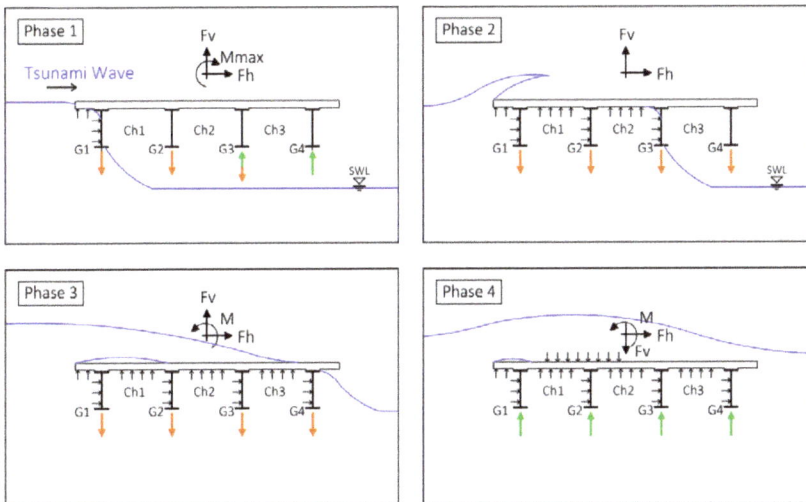

Figure 21. Tsunami inundation mechanism of a bridge.

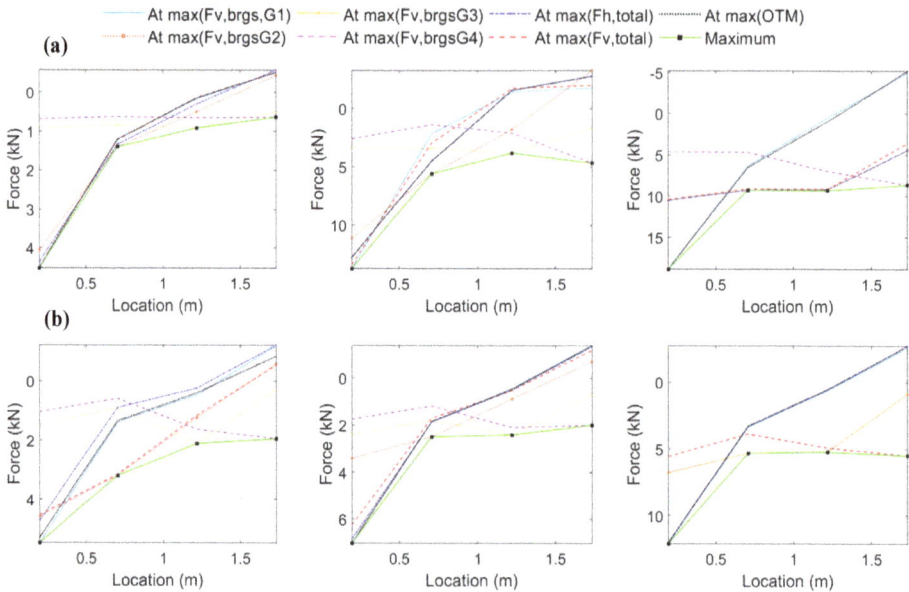

Figure 22. Recorded bearing forces at different instants for (**a**) solitary waves (top) with H = 0.42 m (left), H = 0.55 m (center) and H = 0.70 m (right), and (**b**) bores (bottom) with H = 1.0 m (left), H = 1.10 m (center) and H = 1.40 m (right).

As discussed in the literature review, most of the studies to date have focused on total uplift forces caused by tsunamis and hurricanes. However, the experimental results presented herein are indicating that the total uplift forces alone cannot sufficiently describe the effects on coastal bridges. The reason is the fact that the maximum total uplift can occur either in phase 1, 2 or 3; however, if the maximum occurs in phase 1 then the vertical force is distributed mainly to the four bearings of the first two girders (for the four-girder bridge examined herein), while if it occurs in phase 2 or 3 then the uplift force is distributed to all eight bearings of the girders. This is demonstrated in Figure 22, which depicts the vertical forces recorded in the four bearings at different time-instants. In general, the experimental data reveal that the maximum total uplift force does not necessarily correspond to the maximum uplift forces in all bearings. Therefore, the focus of future research studies should be the forces in the connections and other members of the bridge.

7. New Physics-Based Simplified Methodology for Engineering Practice

One methodology to accurately capture the transient tsunami effects would be the development of time-histories of total horizontal and vertical forces as well as moment. A less sophisticated methodology -presented in Section 5 of this paper- would be the development of tsunami demand diagrams either 2D or 3D ones. This method would be less accurate and economical than developing time-histories but will require less input parameters consequently increasing its practicality. An even simpler methodology would be the determination of several sets of forces with corresponding moments (Fhi, Fvi, OTMi for set i) or with corresponding locations of application (Fhi, locFhi, Fvi, locFvi), which were seen in the experiments to be producing the largest demand on different structural components.

This section will focus on the development of such a simple and practical methodology for predicting the tsunami-induced demand on bearings, columns and connections. To achieve the highest possible accuracy the method will be based on the physics involved in the inundation process of the bridge and all the findings presented in previous sections of this paper. The method will have to be able to simulate the tsunami loading and the generated overturning moment and ideally relate to the 4 phases of the bridge inundation mechanism. Given the fact that the uplift forces in (a) offshore bearings are maximized in Phase 1, (b) bearings G2 are maximized in Phase 1 or 2, (c) bearings G3 are maximized in Phase 2, and (d) bearings G4 are maximized in Phase 2 or 3, the method will have to use different load cases. Last but not least, since the objective of the method is the direct application in engineering practice and previous research studies have already developed equations for the prediction of the maximum total horizontal load (maxFh) and maximum uplift load (maxFv), the ideal case in terms of simplicity would be achieved if the improved method would require only maxFh and maxFv as input parameters. After several iterations and examined approaches, an improved physics-based methodology that meets all the above requirements was developed herein and is shown in Figure 23. The method consists of 3 load cases, which are described below:

- **Load Case 1:** In this case the maximum horizontal force is applied at the mid-height of the offshore girder and the maximum vertical force is applied at the mid-width of the offshore overhang, with the aim to represent Phase 1 of the inundation mechanism and the associated large overturning moment. This load case is expected to give the largest uplift forces in the offshore bearings, columns and connections. The experimental results demonstrated that it is more reasonable to apply the horizontal load at the mid-height of the girder, instead of the mid-height of the bridge or the CG of the deck, since as seen in [23] the recorded pressures histories on the offshore girder at Hgirder/3 and 2Hgirder/3 are simultaneously large in Phase 1, while the pressure on the offshore face of the overhang is maximized much later. It must be clarified though that this is limited only to bridges without barriers (rails) or barriers with perforations that do not have a significant effect on the total horizontal load. This situation might be different for solid barriers and for such a case further investigation is required.

- **Load Case 2:** In this load case the maximum uplift force is applied at the mid-width of chamber 1, while a reduced horizontal load is simultaneously applied at the mid-height of girder G2. The intent for this load case is to capture the effects associated with phase 2 of the inundation mechanism during which the wave could be reaching either chamber 1 or chamber 2. Therefore, the horizontal load is reduced (Fh = a×maxFh) using a calibrated factor that accounts for the observed trends in the recorded time-histories, which showed that for all bore heights and several solitary wave heights the horizontal load was reduced after the initial impact on the offshore girder. A factor a = 0.85 was seen to give reasonable results, however it must be noted that this number is not intended to be conservative in terms of estimating the horizontal force for all solitary waves, since in the experiments some large solitary waves (e.g., H = 0.70 m) showed continuously large horizontal forces during the whole inundation process, so for such waves a = 1 could be used. This load case will generate a smaller overturning moment than load case 1 and could possibly give a more accurate estimation of the uplift forces in bearings G2.

- **Load Case 3:** For load case 3 the maximum uplift force is applied at the mid-width of the onshore chamber 3 together with a reduced horizontal load (Fh = b×maxFh). The objective of this load case is to capture the maximum uplift in bearings G3 and G4, which were seen to be governed by the reverse overturning moment for most wave heights. In this study three values of the reduction factor "b" were examined and particularly b = 0, b = 0.5 and b = 0.65. The zero value would mean that the horizontal load is totally neglected leading to a larger counter-clockwise moment than the one generated by the other two values, and a conservative estimation of the uplift load in bearings G3 and G4.

This physics-based methodology was applied using the experimentally recorded maximum total horizontal force (maxFh) and maximum total uplift force (maxFv) together with Equations (1)–(4) in order to calculate vertical forces in the bearings and the column-bent connections. The ratios of the calculated to the experimentally measured connection forces for Load Cases 1, 2 and 3 are summarized in Tables 5–7 respectively.

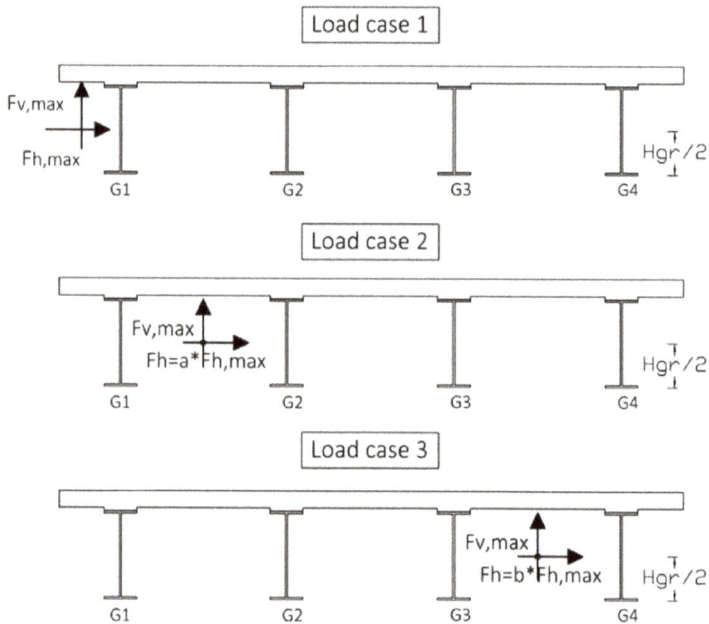

Figure 23. Load cases for the improved physics-based simplified method.

As shown in Table 5, load case 1 estimates uplift forces in the offshore bearings that are conservative for all wave heights apart from two, for which however the predicted force is within 6% from the measured value. For the offshore bearings, this load case gives an over-prediction of 33% on average, demonstrating that the method can conservatively estimate the demand on the offshore connections. For bearings G2 and col. 2 (center) the over-prediction is even more significant with an average value of 66% and 87% respectively. A possible reason for the larger overestimation of the uplift forces in bearings G2 and col. 2 relative to the respective overestimation of the offshore connections, might be the rigid assumption made in Equations (1)–(4), which results in a linear distribution of the uplift force in the connections. On the other hand, Figure 22 showed that the experimentally measured distribution is not linear at the instant the uplift force in the offshore bearings is maximized. Therefore, future work should focus on the development of a 3D numerical model of the bridge that will simulate the actual stiffness of all bridge components and more accurately capture the distribution of the uplift in the connections. Nonetheless, the table shows that load case 1 tends to provide conservative estimates of the uplift demand in bearings G1 and G2 and columns 1 and 2, but fails to capture the respective demand in the rest of the bearings and columns.

The closer location of maxFv to the CG of the bridge in load case 2, and the corresponding smaller overturning moment, is probably the reason this load case fails to predict the uplift in the offshore bearings and columns, but it does give good estimates for bearing G2 and the center column (col. 2). In fact this load case provides a more accurate estimate of the uplift demand in the latter bearing and column than load case 1, by over-predicting on average by 38% and 55%, instead of 66% and 87%, the uplift forces in bearings G2 and col. 2 respectively. Therefore, this load case could be used for the design of bearings G2 and col. 2.

Table 5. Ratios of bearing and column-bent cap connection uplift forces calculated from load case 1 of the improved method to the respective values recorded in the experiments.

Wave Type	Hinput (m)	Brngs, G1	Brngs, G2	Brngs, G3	Brngs, G4	Col. 1	Col. 2	Col. 3
Unbroken solitary	0.46	1.07	1.75	0.24	−3.79	1.12	2.12	−1.87
	0.52	0.94	1.10	0.21	−1.03	1.06	1.58	−0.62
	0.65	1.13	1.59	0.18	−2.07	1.19	1.87	−1.17
	0.36	1.46	1.72	0.01	−1.49	1.47	1.90	−0.82
	0.42	1.10	1.84	0.22	−3.40	1.12	2.43	−1.45
	0.55	0.95	1.23	0.20	−1.16	1.03	1.75	−0.67
	0.70	1.49	1.61	0.19	−1.32	1.53	1.55	−0.92
Bore	0.80	1.18	2.13	−0.06	−2.52	1.21	2.22	−3.79
	1.00	1.56	1.34	0.00	−2.19	1.46	1.87	−2.76
	1.10	1.21	1.67	−0.06	−2.22	1.22	1.69	−2.58
	1.30	1.08	2.05	−0.05	−3.29	1.20	2.03	−3.80
	0.90	1.85	1.72	0.09	−1.28	1.68	1.64	−1.36
	1.00	1.88	1.60	0.08	−1.20	1.76	1.83	−1.39
	1.20	1.22	1.83	0.06	−2.10	1.21	1.82	−2.29
	1.40	1.84	1.76	0.10	−1.51	1.67	1.77	−1.42

Table 6. Ratios of bearing and column-bent cap connection uplift forces calculated from load case 2 of the improved method to the respective values recorded in the experiments.

Wave Type	Hinput (m)	Brngs, G1	Brngs, G2	Brngs, G3	Brngs, G4	Col. 1	Col. 2	Col. 3
Unbroken solitary	0.46	0.79	1.45	0.79	−1.49	0.88	2.12	−0.70
	0.52	0.68	0.91	0.56	−0.35	0.83	1.58	−0.12
	0.65	0.82	1.32	0.55	−0.80	0.93	1.87	−0.45
	0.36	1.09	1.43	0.31	−0.72	1.20	1.90	−0.36
	0.42	0.80	1.52	0.70	−1.32	0.88	2.43	−0.48
	0.55	0.69	1.02	0.51	−0.39	0.80	1.75	−0.15
	0.70	1.08	1.33	0.47	−0.42	1.18	1.55	−0.23
Bore	0.80	0.88	1.77	0.37	−1.29	0.96	2.22	−2.34
	1.00	1.16	1.11	0.35	−1.07	1.17	1.87	−1.54
	1.10	0.90	1.38	0.23	−1.16	0.97	1.69	−1.61
	1.30	0.81	1.70	0.31	−1.69	0.95	2.03	−2.37
	0.90	1.36	1.43	0.38	−0.54	1.31	1.64	−0.64
	1.00	1.39	1.33	0.40	−0.52	1.38	1.83	−0.70
	1.20	0.91	1.52	0.43	−0.96	0.95	1.82	−1.25
	1.40	1.35	1.46	0.41	−0.64	1.31	1.77	−0.69

None of the above two load cases could estimate the uplift in the onshore bearings and columns or in bearings G3, however as shown in Table 7, load case 3 can successfully achieve that. If b = 0 then a conservative estimation of the counter-clockwise moment is made, which results in over-prediction of the uplift forces in bearings G3 and G4 and col. 3 by 52%, 156% and 111% on average respectively. Interestingly, this average value is driven by two of the smaller wave heights with H = 0.42 m and H = 0.46 m (over-prediction of uplift in G4 by 379% and 428%, respectively), which can reach the deck but lose energy after the initial impact on the overhang and offshore girder, and by the time they reach chamber 3 they do not apply significant uplift and consequently counter-clockwise moment. If these two outliers are not included in the calculation the average over-prediction is reduced. An alternative approach to neglecting the total horizontal force (b = 0) is to consider it reduced down to 0.5×maxFh or 0.65×maxFh. The 0.65 factor is an empirical value and is in agreement with trends seen in the experimentally measured bore forces. Although not shown in the table, the 0.65 value yields uplift forces in bearings G3 and G4 that are closer to the measured ones, with the overestimation being 29%, 106% and 25% for bearings G3 and G4 and col. 3 respectively; however, it under-predicts the uplift in the onshore columns for three waves with maximum under-prediction of 35%. Therefore, in order to reduce the under-prediction a value of b = 0.5 was also examined and the results (Table 7) showed an overestimation of 40%, 118% and 45% for bearings G3 and G4 and col. 3 respectively, and an under-prediction of the uplift in the onshore columns only for two waves, which was 13% and 5%.

This demonstrates that engineers could use b = 0.5 for getting more reasonable values of the uplift demand and b = 0 for a conservative design.

Table 7. Ratios of bearing forces calculated from load case 3 of the improved method to the respective values recorded in the experiments.

Wave Type	Hinput (m)	b = 0				b = 0.50			
		Brg1	Brg2	Brg3	Brg4	Brg1	Brg2	Brg3	Brg4
Unbroken solitary	0.46	−0.06	0.57	2.41	5.28	0.01	0.64	2.27	4.68
	0.52	−0.05	0.36	1.54	1.60	−0.01	0.40	1.48	1.48
	0.65	−0.06	0.52	1.65	2.93	0.01	0.58	1.56	2.61
	0.36	−0.07	0.52	1.24	1.67	0.09	0.65	1.11	1.33
	0.42	−0.06	0.60	2.12	4.79	0.01	0.68	2.00	4.26
	0.55	−0.06	0.41	1.40	1.80	−0.01	0.45	1.34	1.67
	0.70	−0.09	0.54	1.25	2.12	−0.03	0.58	1.21	1.99
Bore	0.80	−0.06	0.63	1.72	2.59	0.09	0.81	1.51	1.97
	1.00	−0.08	0.40	1.42	2.42	0.10	0.50	1.27	1.92
	1.10	−0.06	0.48	1.16	2.21	0.10	0.63	1.01	1.65
	1.30	−0.05	0.60	1.43	3.39	0.08	0.78	1.25	2.58
	0.90	−0.10	0.55	1.23	1.68	0.05	0.64	1.15	1.45
	1.00	−0.10	0.50	1.34	1.52	0.06	0.59	1.24	1.29
	1.20	−0.06	0.56	1.57	2.51	0.06	0.68	1.42	2.08
	1.40	−0.10	0.56	1.34	1.97	0.05	0.65	1.24	1.70

It must be noted that as more research studies will be conducted in the future and more information will be generated, the actual values of factors "a" and "b" used in Load Cases 2 and 3 might change so as to result in more accurate estimation of uplift demands and economical designs of structural components. Moreover, some of the assumptions made in the current methodology e.g., the application of maxFv in all three phases, might also be improved and for each load case different magnitudes of the vertical force could be applied. The recommended load cases and factor values are intended for open girder bridges; however, the framework could be adjusted for other types of bridges such as bridges with diaphragms or box-girder bridges, as well as other types of elevated coastal structures such as jetties and wharves or even offshore platforms. Last but not least, it must be clarified that (a) these load cases were seen to give reasonable results for a rigid bridge assumption, however in future studies they will have to be calibrated and used in 3D structural models that will properly simulate the stiffness of all structural components, and (b) the intent of this method is to estimate the uplift demand in individual components and connections assuming that maxFh and maxFv are known parameters. Ongoing and future work of this research team will focus on evaluation, calibration and refinement of predictive equations for maxFh and maxFv.

8. Conclusions

This paper has presented a comprehensive analysis of data obtained during large-scale hydrodynamic experiments of tsunami waves impacting an I-girder bridge with cross-frames. Results obtained have improved the understanding of the tsunami impact and bridge inundation mechanism as well as the associated connection forces between the super- and sub-structures of these bridges. In contrast to the majority of published papers in the field, which focus on maximum total horizontal and vertical forces, this study has focused on forces in individual structural components in order to obtain insight into the temporal and spatial variation of tsunami-induced loading. The main purpose of this paper is to draw attention to bearing, column and connection forces, understand how they are related to wave impact, decipher the inundation process of the deck for unbroken solitary waves and bores, and develop methodologies for the estimation of tsunami demand on components. From the work presented herein it is concluded that:

- Bores introduce a significant, short-duration impulsive (slamming) force followed by a longer duration lateral force with significantly reduced amplitude. This is not the case for unbroken solitary waves. However, both types of waves apply a distinct short-duration impulsive uplift force when the wave hits the offshore overhang followed by (a) a longer duration uplift force as the chambers of the bridge flooded, and (b) a downward force as the deck is inundated.

- The maximum of the horizontal (Fh) and vertical (Fv) forces did not always occur at the same point of the flooding process nor did they coincide in time for all waves, and in fact the horizontal loading was seen to be reduced to 60% of its maximum value at the instant the uplift force was maximized. This demonstrates the transient nature of the tsunami inundation mechanism and its complexity, as well as the need to predict time-consistent vertical and horizontal forces in order to achieve an economical tsunami bridge design.

- The uplift forces in individual bearings and column-bent connections are maximized at different instants of the flooding and overtopping process and have different magnitudes, with the offshore components having to withstand the majority of the total deck uplift, reaching as high as 91% of maxFv for unbroken solitary waves and 96% for bores, with average values being 78% and 70% for the two wave types respectively. Similarly, for the same wave types, the offshore columns are subject to an average of 82% and 94% of maxFv, while for some wave heights the uplift demand can be up to 124% of maxFv, which seems counter-intuitive.

- The maximum uplift force individual bearings and columns do not necessarily coincide with the maximum applied uplift on the deck. This is a major finding because it indicates that the maximum total uplift on the deck might not result in the "worst case" scenario (largest demand) for every structural component, suggesting that the current approach of using the total tsunami force as the sole measure of tsunami demand might not be sufficient.

- The concurrent application of the maximum total horizontal and maximum uplift loading at the CG is not conservative for most bearing and column connections, in contrast to conventional wisdom. The reason behind this observation is the generation of a significant OTM, which has not been fully understood to date. This moment is always maximized at the initial impact of the wave on the offshore girder and overhang, and together with the concurrently large applied uplift force is causing the significant uplift demand on offshore connections and components.

Due to the complex temporal and spatial variation of the tsunami-induced loading, which results in variable effects on individual bearings, it is not possible to identify a priori the combinations of Fh and Fv with corresponding locations of application or corresponding moment that will govern the demand in each connection and structural component. To address this issue, the paper presents "tsunami demand diagrams", which are 2D envelopes of (Fh, Fv) and (OTM, Fv), and 3D envelopes of (Fh, Fv, OTM). The demand diagrams together with the time-histories revealed significant differences in the patterns of horizontal and vertical forces, as well as the overturning moment between the unbroken solitary waves and bores, indicating that the physics governing the wave impact and induced effects on the bridge are different for the two wave types. This suggests the need to develop methodologies that will be able to predict the wave type expected at a specific bridge location, in order to design the connections safely and without undue conservatism.

Another topic of investigation in this paper was the tsunami inundation mechanism of bridges, which consisted of four main phases, from the time of the initial impact to the full overtopping of the bridge. Among these four phases, there exists a (a) short duration phase with large impulsive horizontal and vertical loads, and occurs when the wave hits the offshore girder and is trapped momentarily under the offshore overhang, generating the maximum horizontal bore forces, the maximum overturning moment and maximum uplift in the offshore bearings and column connections; (b) a longer duration phase that starts when the wave reaches chamber 1 and finishes when the wave has flooded the center chamber of the bridge, at which point the OTM is small but the total deck uplift force is maximized (for several waves) and is distributed to all bearings and all columns, governing the design of the center ones, and (c) a phase that occurs when the wave reaches the onshore chamber, at which point the uplift load is large and significant counter-clockwise moment is generated, introducing the largest uplift force in the onshore bearings, columns and connections for most waves. Based on the bridge inundation mechanism, a new physics-based methodology was developed for providing conservative estimates of uplift demand in each connection. The practicality of this methodology lies in the fact that it uses as an input only (a) two parameters -maxFh and maxFv- for which several predictive equations are available in the literature, and (b) two factors that have been calibrated based on the experimental results presented herein.

The findings in this paper indicate the need for a paradigm shift in the assessment of tsunami risk to coastal bridges to include not just the estimation of total tsunami load on a bridge but also the distribution of this load to individual structural components that are necessary for the survival of the bridge. It is acknowledged though that the findings presented in this paper are limited to bridges with four open girder lines with cross-frames, eight steel bearings, and no hand rails or crash barriers. Moreover, the findings might be limited by the assumptions made in the simplified methodology and particularly the fact that (a) the deck is assumed to be rigid, (b) the reactive moments in the steel bearings and in the column-bent connections are assumed to be zero, and (c) inertia and damping effects are not considered. Future studies should address these limitations and could attempt to expand the methodologies presented herein to other types of bridges or coastal decks, such as jetties.

Author Contributions: I.G.B. and D.I. conceived the need to more rigorously examine the hydraulic demand on connections and components in bridge superstructures during tsunami overtopping than had been done in the past, and designed and executed the large-scale experiments described in this paper. D.I. conducted the in-depth analyses of the experimental data, developed the technical content presented in this paper and wrote the first draft, while I.G.B. provided advice, feedback and editing of the paper. P.L. and S.Y. assisted with the design and conduct of the experiments, reviewed the current manuscript, and provided feedback.

Funding: The hydrodynamic experiments and a portion of the analyses presented in this paper were funded by the Federal Highway Administration (FHWA) under Contract No. DTFH61-07-C-00031, "Improving the Seismic Resilience of the Federal-Aid Highway System", awarded to Ian Buckle, PI. The development of the physics-based methodology for estimating connections forces and the preparation of this manuscript was funded by the Oregon Department of Transportation (ODOT) under Contract No. 32399, "Verification of Tsunami Bridge Design Equations" awarded to Ian Buckle, PI, and Denis Istrati, Co-PI.

Acknowledgments: During the FHWA-funded project, technical guidance was provided by the following FHWA Representatives of the Contracting Officer (COR): Wen-huei (Phillip) Yen, Fred Faridazar and Sheila Duwadi. During the ODOT-funded project, research coordination was provided by Jon Lazarus and technical guidance was given by Bruce Johnson and Albert Nako from ODOT. The authors also recognize Tao Xiang, Anastasia Bitsani, Patrick Laplace, Chad Lyttle, Todd Lyttle, Tim Maddux, Alicia Lyman-Holt, James Batti, Jeff Gent for their technical assistance during the experiments. Moreover, the authors wish to thank Daniel Cox and Chris Higgins for agreeing to the author's modification and upgrade of their experimental setup so it could be used for the tsunami experiments described in this paper. Last but not least, the authors appreciate the assistance provided by Information Technology at UNR and access to the HPC resources necessary to run the advanced CFD and FSI numerical analyses for the hydrodynamic design of the experiments.

Conflicts of Interest: The authors declare no conflict of interest.

References

1. Unjoh, S. Bridge damage caused by tsunami. *Jpn. Assoc. Earthq. Eng.* **2007**, *6*, 6–28.
2. Maruyama, K.; Tanaka, Y.; Kosa, K.; Hosoda, A.; Arikawa, T.; Mizutani, N.; Nakamura, T. Evaluation of tsunami force acting on bridge girders. In Proceedings of the Thirteenth East Asia-Pacific Conference on Structural Engineering and Construction (EASEC-13), Sapporo, Japan, 11–13 September 2013; Keynote-Lecture.
3. Kosa, K. Damage analysis of bridges affected by tsunami due to Great East Japan Earthquake. In Proceedings of the Symposium on Engineering Lessons Learned from the 2011 Great East Japan Earthquake, Tokyo, Japan, 1–4 March 2012.
4. Kasano, H.; Oka, J.; Sakurai, J.; Kodama, N.; Yoda, T. Investigative research on bridges subjected to tsunami disaster in 2011 off the pacific coast of Tohoku earthquake. In *Australasian Structural Engineering Conference 2012: The Past, Present and Future of Structural Engineering*; Engineers Australia: Barton, Australia, 2012; p. 51.

5. Kawashima, K. Damage of bridges due to the 2011 great east japan earthquake. In Proceedings of the International Symposium on Engineering Lessons Learned from the 2011 Great East Japan Earthquake, Tokyo, Japan, 1–4 March 2012.

6. Lau, T.L.; Ohmachi, T.; Inoue, S.; Lukkunaprasit, P. Experimental and numerical modeling of tsunami force on bridge decks. In *Tsunami-a Growing Disaster*; InTech: London, UK, 2011.

7. Hayashi, H. Study on tsunami wave force acting on a bridge superstructure. In Proceedings of the 29th US-Japan Bridge Engineering Workshop, Tsukuba, Japan, 11–13 November 2013.

8. Seiffert, B.; Hayatdavoodi, M.; Ertekin, R.C. Experiments and computations of solitary-wave forces on a coastal-bridge deck. Part I: Flat plate. *Coast. Eng.* **2014**, *88*, 194–209. [CrossRef]

9. Hayatdavoodi, M.; Seiffert, B.; Ertekin, R.C. Experiments and computations of solitary-wave forces on a coastal-bridge deck. Part II: Deck with girders. *Coast. Eng.* **2014**, *88*, 210–228. [CrossRef]

10. Bozorgnia, M.; Lee, J.J.; Raichlen, F. Wave structure interaction: Role of entrapped air on wave impact and uplift forces. *Coast. Eng. Proc.* **2011**, *1*, 57. [CrossRef]

11. Yim, S.C.; Boon-intra, S.; Nimmala, S.B.; Winston, H.M.; Azadbakht, M.; Cheung, K.F. *Development of a Guideline for Estimating Tsunami Forces on Bridge Superstructures*; Report No. OR-RD-12-03; Oregon Department of Transportation: Salem, OR, USA, 2011.

12. Kataoka, S.; Kaneko, M. Estimation of wave force acting on bridge superstructures due to the 2011 Tohoku Tsunami. *J. Disaster Res.* **2013**, *8*, 605–611. [CrossRef]

13. Nakao, H.; Zhang, G.; Sumimura, T.; Hoshikuma, J. Numerical assessment of tsunami-induced effect on bridge behavior. In Proceedings of the 29th US-Japan Bridge Engineering Workshop, Tsukuba, Japan, 11–13 November 2013.

14. Bricker, J.D.; Nakayama, A. Contribution of trapped air deck superelevation, and nearby structures to bridge deck failure during a tsunami. *J. Hydraul. Eng.* **2014**, *140*, 05014002. [CrossRef]

15. Azadbakht, M.; Yim, S.C. Simulation and estimation of tsunami loads on bridge superstructures. *J. Waterw. Port Coast. Ocean Eng.* **2014**, *141*, 04014031. [CrossRef]

16. Istrati, D.; Buckle, I.G. Effect of fluid-structure interaction on connection forces in bridges due to tsunami loads. In Proceedings of the 30th US-Japan Bridge Engineering Workshop, Washington, DC, USA, 21–23 October 2014.

17. Fu, L.; Kosa, K.; Sasaki, T. Tsunami damage evaluation of utatsu bridge by video and 2-d simulation analyses. *J. Struct. Eng.* **2013**, *59*, 428–438.

18. Kawashima, K.; Buckle, I. Structural performance of bridges in the tohoku-oki earthquake. *Earthq. Spectra* **2013**, *29*, S315–S338. [CrossRef]

19. Araki, S.; Ishino, K.; Deguchi, I. Stability of girder bridge against tsunami fluid force. *Coast. Eng. Proc.* **2011**, *1*, 56. [CrossRef]

20. Rahman, S.; Akib, S.; Shirazi, S.M. Experimental investigation on the stability of bride girder against tsunami forces. *Sci. China Technol. Sci.* **2014**, *57*, 2028–2036. [CrossRef]

21. Maruyama, K.; Hosoda, A.; Tanaka, Y.; Kosa, K.; Arikawa, T.; Mizutani, N. Tsunami force acting on bridge girders. *J. JSCE* **2017**, *5*, 157–169. [CrossRef]

22. Mazinani, I.; Ismail, Z.; Hashim, A.M.; Saba, A. Experimental investigation on tsunami acting on bridges. *Int. J. Civ. Archit. Struct. Constr. Eng.* **2014**, *8*, 1040–1043.

23. Istrati, D. Large-Scale Experiments of Tsunami Inundation of Bridges Including Fluid-Structure-Interaction. Doctoral Dissertation, University of Nevada, Reno, NV, USA, May 2017.

24. Hoshikuma, J.; Zhang, G.; Nakao, H.; Sumimura, T. Tsunami-induced effects on girder bridges. In Proceedings of the International Symposium for Bridge Earthquake Engineering in Honor of Retirement of Professor Kazuhiko Kawashima, Tokyo, Japan, 15 March 2013.

25. Martinelli, L.; Lamberti, A.; Gaeta, M.G.; Tirindelli, M.; Alderson, J.; Schimmels, S. Wave loads on exposed jetties: Description of large-scale experiments and preliminary results. In Proceedings of the Conference on Coastal Engineering, Shanghai, China, 30 June–5 July 2010.

26. Murakami, K.; Sakamoto, Y.; Nonaka, T. Analytical investigation of slab bridge damages caused by tsunami flow. *Coast. Eng. Proc.* **2012**, *1*, 42. [CrossRef]

27. Motley, M.R.; Wong, H.K.; Qin, X.; Winter, A.O.; Eberhard, M.O. Tsunami-induced forces on skewed bridges. *J. Waterw. Port Coast. Ocean Eng.* **2015**, *142*, 04015025. [CrossRef]

J. Mar. Sci. Eng. **2018**, *6*, 148

28. Wei, Z.; Dalrymple, R.A. Numerical study on mitigating tsunami force on bridges by an SPH model. *J. Ocean Eng. Mar. Energy* **2016**, *2*, 365–380. [CrossRef]

29. Zhu, M.; Elkhetali, I.; Scott, M.H. Validation of opensees for tsunami loading on bridge superstructures. *J. Bridge Eng.* **2018**, *23*, 04018015. [CrossRef]

30. Xu, G.; Cai, C.S.; Hu, P.; Dong, Z. Component level–based assessment of the solitary wave forces on a typical coastal bridge deck and the countermeasure of air venting holes. *Pract. Period. Struct. Des. Constr.* **2016**, *21*, 04016012. [CrossRef]

31. Winter, A.O.; Motley, M.R.; Eberhard, M.O. Tsunami-like wave loading of individual bridge components. *J. Bridge Eng.* **2017**, *23*, 04017137. [CrossRef]

32. Istrati, D.; Buckle, I.G.; Itani, A.; Lomonaco, P.; Yim, S. Large-scale fsi experiments on tsunami-induced forces in bridges. In Proceedings of the 16th World Conference on Earthquake Engineering, Santiago, Chile, 9–13 January 2017.

33. Douglass, S.L.; Chen, Q.; Olsen, J.M.; Edge, B.L.; Brown, D. *Wave Forces on Bridge Decks*; Final Report Prepared for U.S.; Department of Transportation and Federal Highway Administration Office of Bridge Technology: Washington, DC, USA, 2006.

34. McPherson, R.L. Hurricane Induced Wave and Surge Forces on Bridge Decks. Master's Thesis, Texas A&M University, College Station, TX, USA, 2008.

35. Bradner, C.; Schumacher, T.; Cox, D.; Higgins, C. Experimental setup for a large-scale bridge superstructure model subjected to waves. *J. Waterw. Port Coast. Ocean Eng.* **2010**, *137*, 3–11. [CrossRef]

36. Hayatdavoodi, M.; Ertekin, R.C. Review of wave loads on coastal bridge decks. *Appl. Mech. Rev.* **2016**, *68*, 030802. [CrossRef]

37. AASHTO. *Guide Specifications for Bridges Vulnerable to Coastal Storms*; AASHTO: Washington, DC, USA, 2008.

38. FEMA P-646. *Guidelines for Design of Structures for Vertical Evacuation from Tsunamis*, 2nd ed.; Federal Emergency Management Agency: Washington, DC, USA, 2012; 174p.

39. *MATLAB User's Guide Release 2016a*; The MathWorks, Inc.: Natick, MA, USA, 2016.

© 2018 by the authors. Licensee MDPI, Basel, Switzerland. This article is an open access article distributed under the terms and conditions of the Creative Commons Attribution (CC BY) license (http://creativecommons.org/licenses/by/4.0/).

Journal of
Marine Science and Engineering

MDPI

Article

Experimental Study on Extreme Hydrodynamic Loading on Pipelines. Part 1: Flow Hydrodynamics

Behnaz Ghodoosipour [1,*], Jacob Stolle [1], Ioan Nistor [1], Abdolmajid Mohammadian [1] and Nils Goseberg [2]

[1] Department of Civil Engineering, University of Ottawa, Ottawa, ON K1N 6N5, Canada
[2] Leichtweiß-Institute for Hydraulic Engineering and Water Resources, Technische Universität Braunschweig, 38106 Braunschweig, Germany
* Correspondence: bghod068@uottawa.ca; Tel.: +1-613-562-5800

Received: 23 May 2019; Accepted: 24 July 2019; Published: 31 July 2019

Abstract: Over the past two decades, extreme flood events generated by tsunamis or hurricanes have caused massive damage to nearshore infrastructures and coastal communities. Utility pipelines are part of such infrastructure and need to be protected against potential extreme hydrodynamic loading. Therefore, to address the uncertainties and parameters involved in extreme hydrodynamic loading on pipelines, a comprehensive experimental program was performed using an experimental facility which is capable of generating significant hydraulic forcing, such as dam-break waves. The study presented herein examines the dam-break flow characteristics and influence of the presence of pipelines on flow conditions. To simulate conditions of coastal flooding under tsunami-induced inundation, experiments were performed on both dry and wet bed conditions to assess the influence of different impoundment depths and still water levels on the hydrodynamic features.

Keywords: pipelines; extreme events; tsunami; dam-break wave; hydrodynamics

---

1. Introduction

1.1. Background

Recent devastating tsunami and storm surge events exposed the vulnerability of coastal communities to such extreme natural disasters. The number of people experiencing such catastrophic coastal flood events has been compounded by climate change and the ever-increasing urbanization of low-lying coastal areas all around the world [1]. This provided increased interest for research around the topic of extreme impacts on infrastructure. The need to study the hydrodynamic loading induced by such events and its effects on various structures is important. Coastal-induced inundation, due to tsunamis, hurricanes, and associated storm surges, can generate extreme turbulence, which impacts coastal areas and destroys infrastructures in their path. Moreover, sudden dam failure incidents can also cause similar impacts to vulnerable downstream infrastructures [2]. Understanding the dynamics of highly turbulent waves, and transient flows, as well as their interaction with structures, is complex and difficult to assess and quantify. This is among others, due to their highly non-linear and rapidly transient characteristics, and the common involvement of turbulent multi-phase processes [3]. Several researchers have conducted post-event forensic field surveys of the recent catastrophic events, such as the 2004 Indian Ocean and the 2011 Tōhoku Tsunami. Field survey results after the 2004 Indian Ocean Tsunami conducted in Khao Lak, Thailand, estimated coastal inundation heights between 4 to 7 m [4,5] and wave front celerities between 6 and 8 m/s [6]. Data recorded by the Japanese Port and Airport Research Institute (PARI) after the 2011 Tōhoku Tsunami in Japan showed inundation depths of up to 15 m in the city of Onagawa. During the same event, onshore inundation velocities of up to 10 to 13 m/s were also observed near Sendai Airport [7]. Fritz et al. [8] used video-processing of images filmed in Kasennuma Bay during the 2011 Tōhoku Tsunami and estimated inundation depths up to 9 m, flow velocities of up to 11 m/s and calculated Froude number values around 1. Field and numerical modelling of the 1993 Hokkaido-Nansei-Oki Tsunami revealed water depths of 5–15 m and flow velocities of 3–15 m/s. Results from such detailed surveys provide invaluable sources of hydrodynamics data for further analysis and comparison with experimental data or available analytical or empirical formula.

Several studies attempted to characterize the hydrodynamics of tsunami run-up on coastlines and its interaction with structures subjected to a solitary wave as a representative of the tsunami [9–12]. Aristodemo et al. [13] conducted a small-scale experimental study in a wave flume along with a numerical investigation using a smoothed-particle hydrodynamics (SPH) model. In their study, they investigated the induced loading from a solitary wave on a horizontal circular cylinder. Although solitary waves have been extensively used for tsunami-related studies, more in-depth studies have shown that such waves cannot represent real tsunamis properly, due to the discrepancies, such as difference in wave period, and height during the wave run-up [14]. Madsen et al. [15] concluded that the required evolutionary distance for an initial run-up into a solitary wave is well beyond the width of any oceanic dimension, concluding that solitary waves would not represent real tsunamis. Chan et al. [16], referred to the available data from the 2011 Tōhoku tsunami and concluded that the wavelength for real tsunami is significantly longer than for solitary waves generated in the laboratory.

Several studies tried to compare the different characteristics of solitary waves and a more realistic representative for tsunami-like waves, such as "bores". Leschka and Oumeraci [17] investigated the induced hydrodynamic forces from two different types of waves, namely solitary waves and bores, representing tsunamis, on three vertical cylinders with different arrangements numerically. They concluded that two different waves result in different flow hydrodynamics, i.e., wave height and flow velocity. Istrati et al. [18] investigated the different types of tsunamis, i.e., solitary waves and bores, and their effect on I-grider bridge with cross-frames. They characterized the induced forces from a bore as short-duration impulsive horizontal force at the time of bore impact that is followed by a smaller magnitude and longer duration forces. This was not observed in the case of solitary waves. Such findings emphasize the importance of choosing the correct type of tsunami representation for characterizing the induced forces on structures, especially for long waves. They also indicate that the unbroken solitary waves are not suitable wave models representing tsunamis.

Zhao et al. [19] studied the hydrodynamic properties of submarine pipelines under the impact of several widely used waves representing tsunamis, including solitary waves and N-waves with characteristics closer to a real tsunami. This study suggests that the hydrodynamic characteristics of the waves, such as water level, flow velocity, flow structure and induced forces in these methods are largely different. Moreover, the longer periods in tsunami-like waves causes a smoother water surface profile compared to solitary waves with shorter periods. As the wave passes the pipe, the size of the vortices generated downstream of the pipe in a solitary-wave is smaller than the vortices generated by tsunami-like waves. Moreover, by the time the wave passes the pipe, the wave height decreases faster for solitary waves as compared to tsunami-like waves, which in turn reduces the induced forces, as well as the duration of the acting force.

In summary, in this study, dam-break waves were used for studying the tsunami-like impact on pipes. Several researchers, i.e., References [20,21], have characterized such waves and stated that dam-break waves could represent real tsunamis.

In this study, where the assessment of tsunami-induced coastal floods on pipelines is addressed, a dam-break wave generated using a rapidly-opening swing gate was used to reproduce the highly turbulent flow conditions created during such extreme events. Stolle et al. [22] described details of the discussed dam-break waves. Several researchers characterized dam-break waves surging over a dry bed. Among them, Ritter [23], Henderson [24] and Chanson [25] developed solutions for the dam-break wave profile. The effect of bed condition (i.e., dry and wet bed) has also been the subject of several studies. Chanson et al. [20] performed an experimental study on tsunami characteristics on wet and dry horizontal beds. They characterized wave momentum and wave front velocity at the beginning of the wave propagation in tsunamis and compared them to the classical dam-break waves. Wuthrich et al. [26] proposed a new method to generate bores over dry and wet bed conditions and investigated the influence of different parameters of the wave, such as the bore front celerity and the flow velocity profile. Moreover, studies, such as St. German et al. [27] and Douglas and Nistor [28] have investigated the effect of bed condition on tsunami characteristics numerically.

Several other researchers investigated the impact of dam-break and tsunami on structures. Nouri et al. [29], Al-Faesly et al. [30], Bremm et al. [31] and Foster et al. [32] evaluated the forces induced from tsunami-structure interaction under unsteady conditions. Other studies such as Wüthrich et al. [33,34] investigated the extreme hydrodynamic forces induced on buildings with various characteristics. Arnason [35] studied the interaction between an incident bore and a free-standing structure and focused on analyzing flow hydrodynamics in their presence. Goseberg et al. [36], studied different flow characteristics around vertical obstacles impacted by transient tsunami-like long waves. Studies, such as Araki and Deguchi [37], Mazinani et al. [38], and Chen et al. [39], investigated tsunami bore impact on coastal bridges.

An in-depth review of existing research in the context of extreme flow condition impact on different structures reveals a lack of investigations on the impact of transient tsunami-like waves on horizontal pipelines. The American Society of Civil Engineers (ASCE), through its ASCE7 Tsunami Loads and Effects Committee, has developed a new standard for tsunami impacts and loading [40]. Amongst the potential effects of such extreme events on infrastructure, this standard has emphasized the need to investigate tsunami loads on pipelines. The work presented herein is the first of a two-part paper which focuses for the first time, to the knowledge of the authors, on the impact of tsunami-like dam-break waves on submerged and above-ground horizontal pipelines in on wet and dry bed conditions. Part 1, focuses on the free-stream flow hydrodynamics and its alterations in the presence of a pipe impacted by such a flow. Part 2 [41] focuses on the hydrodynamic loading and associated force coefficients for horizontal pipelines located in coastal areas prone to tsunamis.

1.2. Objectives

The main goal of this study is to investigate the extreme hydrodynamic loading on pipelines under transient flows. The specific objectives of this first part of the two-part paper are to investigate and discuss the flow hydrodynamics of the dam-break waves and movement and impingement on a pipeline installed in its flow path. The following specific questions were examined in this study:

- What are the flow characteristics (time-history of the wave surface profile and flow velocity) for dam-break waves propagating over dry bed conditions for different wave heights?
- How are flow characteristics altered in the case of dam-break wave propagation over wet bed (still water on the flume bed, downstream of the impounding gate) and how are these characteristics changing when the dam-break wave height changes and/or when the still water depth of the wet bed varies?
- How do flow conditions get influenced by the presence of a horizontal cylindrical pipe immersed in the flow under both dry and wet bed conditions?

Results from this first paper will be further used to analyze and discuss the complex behavior of the hydrodynamic loading exerted on the pipe: In the companion paper [41].

2. Experimental Setup

2.1. Dam-Break Flume

A comprehensive experimental program was developed and conducted in the Dam-break Flume in the Hydraulics Laboratory at the University of Ottawa, Canada. Experiments were performed at a 1:25 length scale, under Froude similitude. The flume is 30.1 m long, 1.5 m wide, and 0.5 m high. A predetermined volume of water was impounded behind a rapidly-opening swing gate installed in the flume to form an upstream reservoir with a length of 21.55 m and variable water depths. The dam-break waves were generated by the rapid opening of the swing gate. According to a previous study conducted in the same flume by Stolle et al. [22], the non-dimensional gate opening time, $T_0 = (t\sqrt{g/h})$ (t being the gate opening time, h being the impoundment depth), is dependent on h with an approximately linear relationship as:

$$T_0 = 1.47\text{--}1.19\,h. \tag{1}$$

With the range of impoundment depths tested in this study, the non-dimensional gate opening time was in the range of $0.875 < T_0 < 1.113$. Therefore, the range of T_0, satisfied the criterion for acceptable non-dimensional gate opening time $T_0 < 1.41$ defined by Lauber and Hager [42]. A vertically-moving steel gate located at the downstream end of the flume ensured the control of the water level downstream of the swing gate and enabled, thus, the formation of wet bed conditions with different water depths. The tested pipe model was made of cold-rolled steel, 10 cm in diameter and 1.47 m in length, installed horizontally and transversally across the flume at $x = 5.6$ m downstream of the gate. The pipe was designed to be perfectly rigid with very high natural frequency in order to avoid possible dynamic effects. Schematics of the flume, together with various experimental parameters, are shown in Figure 1.

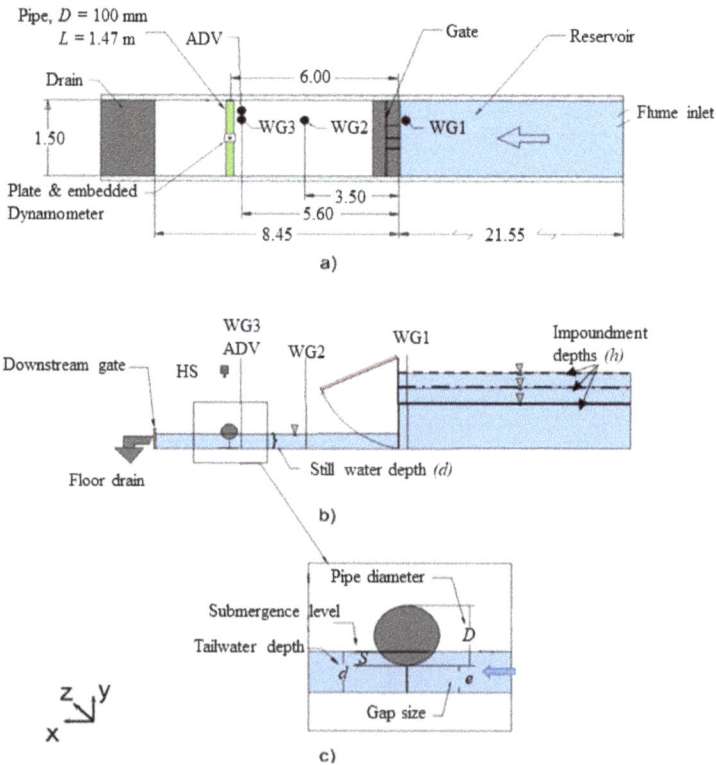

Figure 1. Flume and instrumentation sketch, (unless otherwise specified, all dimensions are in m). (**a**) Plan view, (**b**) side view and (**c**) close view with pipe and experimental parameters.

2.2. Instrumentation

Figure 2a,b show images of the flume together with the instruments and their locations. $x = 0$ was defined at the longitudinal position of the gate.

Figure 2. (**a**) Downstream view of pipe, dynamometer, ADV and wave gauge, (**b**) downstream view of flume and gate and (**c**) close view of pipe, supports and the base plate.

2.2.1. Wave Gauges

To record the time-history of the water level, three wave gauges (RBR WG-50, capacitance-type, ±0.002 m accuracy, RBR Global, Ottawa, ON, Canada) were installed at different locations along the flume. The first wave gauge (WG3) was installed upstream of the gate ($x = -0.04$ m) and was used to determine the opening time of the gate. The instant when the reservoir water level started to decrease, which was recorded by WG1, was used to synchronize the other measurement instruments The other two wave gauges were located at $x = 3.5$ m (WG2) and $x = 5.6$ m (WG1) downstream of the gate. The wave gauges sampling rate was 300 Hz. Wave gauges were calibrated by ensuring a linear gauge response with R^2 values greater than 0.99.

2.2.2. Acoustic Doppler Velocimeter (ADV)

A high-resolution acoustic Doppler velocimeter (ADV) (Vectrino, ±1 mm/s accuracy, 2.5 m/s measurement range, Nortek, Norway) was used for velocity measurements in the free stream flow. The velocity was used in the estimation of the drag and lift coefficients. The ADV was able to measure 3-D water velocities using coherent Doppler processing technology. In this study, a side looking ADV was used. The ADV's sampling rate was set to 200 Hz. The instrument was located at $x = 5.6$ m, 0.10 m upstream of the outer edge of the pipe. To derive the velocity profile, each experiment was repeated three times, and the ADV was moved vertically to different depths: (1) The highest water level; (2) the location where the center-axis of the pipe cross section was placed and 0.03 m above the flume bed. Non-uniformities in the cross-flow direction were assumed to be negligible. The water was seeded before each test using aluminum oxide powder with 27 micron particle size (400 mesh) to ensure adequate signal to noise ratio for ADV measurements.

2.2.3. Dynamometer

To record the time-history of the forces exerted on the pipe, a 6 degree of freedom (DOF) dynamometer (Interface- 6A68E, non-linearity, 0.04%, maximum capacity: $Fx = Fy = 10$ kN, $Fz = 20$ kN, $Mx = My = Mz = 500$ Nm, Interface Inc., Scottsdale, AZ, USA) was used. This dynamometer was able to simultaneously measure the time-histories of the forces and moments along the three axes. The dynamometer was installed beneath the concrete flume floor by cutting the concrete flume floor, placing the device and re-embedding it, as shown in Figure 3a,b.

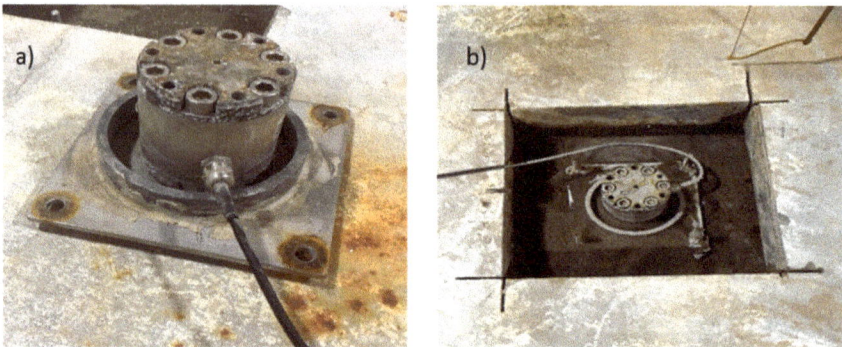

Figure 3. (**a**) Dynamometer (Interface-6A68E) and (**b**) dynamometer embedded in the flume floor.

A stiff steel plate was placed on top of the instrument, levelled with the flume. This plate transmitted the exerted force to the dynamometer through four bolts which rigidly fastened the instrument to the pipe model. The cable was securely placed in the flume floor recess, ensuring that no additional forces were transmitted on to the transducer. The pipe was attached to the plate using two very narrow, vertical plates, as shown in Figure 2a. The dynamometer was calibrated using the calibration chart provided by the manufacturer and re-zeroed before each test. The sampling rate for the dynamometer was set to 300 Hz. The amount of force and moment applied to the dynamometer was converted to voltage and recorded by the data acquisition system.

2.2.4. Data Acquisition System

Analog voltage signals from different instruments used in the experiments were converted to digital format and saved into data files using the QuantumX data acquisition system (MX840B, 8-channel universal amplifier and MX1601B with 16 individually configurable channels, HBM, Marlborough, MA, USA). All data were synchronized between the devices using a FireWire connection.

2.2.5. Camera

A camera (HS, Flare 2M360-CL, sampling rate 70 Hz, IO Industrial, London, ON, Canada) was directed towards the pipe from top to capture and analyze the bore impact with the pipe. A GoPro Hero4 Black (GoPro, San Mateo, CA, USA) was also installed 2 m upstream of the pipe and was used for observation purposes.

2.2.6. Cylindrical Pipe

A steel pipe, referred to as the cylindrical pipe, with an outer diameter of 100 mm, a wall thickness of 5 mm and a length of 1470 mm was used in the experiments. The pipe was connected to the upper plate bolted to the dynamometer using two brackets, 2 mm thick, made of steel, as shown in Figure 1a.

2.3. Experimental Test Program

Findings from this study were used in the companion paper (part 2) to characterize hydrodynamic forces exerted on pipelines, due to extreme flow events, modelled using a dam-break wave. A systematic and comprehensive experimental approach was conducted for this purpose. The most relevant parameters governing the problem at hand were varied during the experiments, namely: Reservoir depth (h), tailwater depth (d), lower edge of pipe distance to bed (e) to diameter ratio (e/D) and pipe level of submergence (S) to pipe diameter ratio (S/D). (Figure 1c). Table 1 shows the list of hydrodynamic tests in the absence of the pipe, while Table 2 illustrates the list of experiments in the presence of a pipe with different experimental configurations. Each test was repeated three times to assess the repeatability of the results of each test. Head ratio (d/h) is defined as the ratio between the still water depth (d) and impoundment depth (h).

Table 1. List of hydrodynamic tests (no pipe).

	Reservoir Depth h (m)	Still Water Depth d (m)	Head Ratio d/h (-)
Hydrodynamic test (no pipe)	0.3	0	0
		0.03	0.1
		0.06	0.2
		0.08	0.26
		0.12	0.4
		0.17	0.56
Hydrodynamic test (no pipe)	0.4	0	0
		0.03	0.075
		0.06	0.15
		0.08	0.2
		0.12	0.3
		0.17	0.425
Hydrodynamic test (no pipe)	0.5	0	0
		0.03	0.06
		0.06	0.12
		0.08	0.16
		0.12	0.24
		0.17	0.34

Table 2. List of experimental configurations in the presence of a pipe.

Gap Ratio e/D (-)	Reservoir Depth h (m)	Still Water Depth d (m)	Head Ratio d/h (-)	Level of Submergence Ratio S/D (-)
0.3	0.3	0	0	0
	0.40	0.03	0.1	0
	0.50	0.06	0.2	0.3
0.6	0.30	0.08	0.26	0.5
	0.40	0.12	0.4	1
	0.4	0	0	0
0.8	0.30	0.03	0.075	0
	0.40	0.06	0.15	0.3
	0.50	0.08	0.2	0.5

Test Repeatability (Water Level Time History)

Multiple tests with identical impoundment depths and initial still water depths were carried out to verify the repeatability of the tests. Figure 4 shows the water surface profile in dry and wet bed conditions measured by WG2 at $x = 3.5$ m. Good agreement of water surface time-histories was achieved between multiple repetitions for both dry and wet bed conditions. Normalized standard deviations (σ/h) of less than 5% for wet bed and less than 4% for the dry bed were obtained.

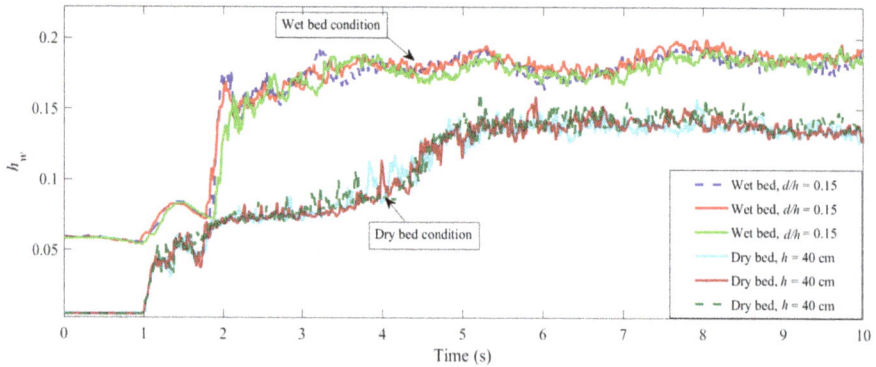

Figure 4. Repeatability of tests for water level time history (WG2), with various water depth (*h*) values of dry bed condition and *d/h* values of the wet bed condition.

3. Results and Discussion

3.1. Dry Bed Condition Hydrodynamics

3.1.1. Dry Bed Water Surface Profile

Figure 5 illustrates the measured water surface profile at the locations of the three wave gauges, shown in Figure 1. Figure 5a shows the normalized water level h_w/h time-history at the location of the reservoir wave gauge (WG1), indicating a decrease in reservoir water depth in time. The gate opening time was used as the reference time in all three figures. Figure 5b,c illustrate the water surface profile for WG2 at $x = 3.5$ m and WG3 at $x = 5.5$ m (at the pipe location) downstream of the gate, respectively. Both Figure 5b,c illustrate the earlier wave arrival time for the dry bed surges generated using larger impoundment depth indicating a larger bore front celerity. The bore front celerity is discussed in more detail in Section 3.1.2. Comparison between water level magnitudes in Figure 5b,c also indicates a decrease in water level as the surge moves forward through the flume. The average non-dimensional water level ($\frac{h_w}{h}$) was decreased by 32% at WG3 compared to WG2. This can be explained with energy losses, due to bed friction in the case of flow on dry bed condition.

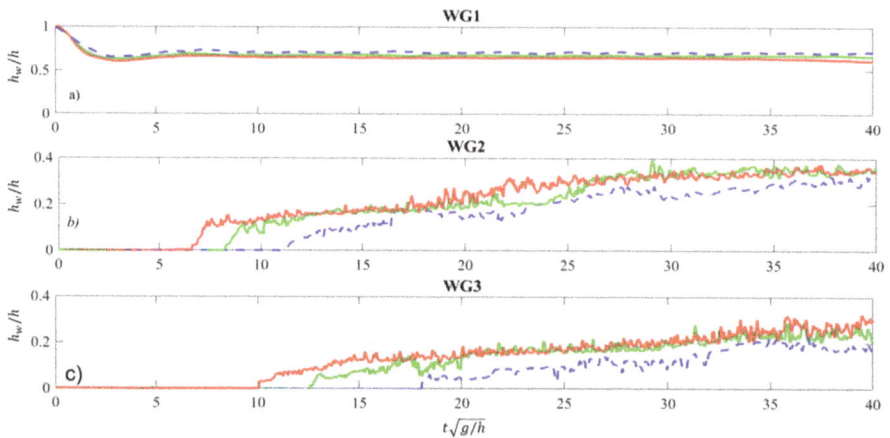

Figure 5. Dry bed surge time-history of the water surface profile. $h = 50, 40, 30$ cm measured at location of the (**a**) reservoir wave gauge (WG1), (**b**) $x = 3.5$ m (WG2) and (**c**) $x= 5.5$ m (WG3).

Figure 6 compares experimental results for the water surface profiles for dam-break flow in dry bed conditions, obtained from multiple tests with identical gate opening conditions and wave gauge locations along the flume, with the analytical solution of the Saint-Venant equations for a horizontal, frictionless surface given by Ritter [23] as:

$$\frac{h_w}{h} = \frac{1}{9}\left(2 - \frac{x}{ht\sqrt{\frac{g}{h}}}\right)^2 \tag{2}$$

In Figure 6, water surface profiles are plotted versus the dimensionless time, $t\sqrt{g/h}$. Experimental and Ritter's theoretical solution were compared at WG2 (x = 3.5 m) for three different impoundment depths h = 30, 40, and 50 cm. Figure 6 shows that, initially ($0 < t\sqrt{g/h} < 2$), the experimental results do not accurately match the Ritter [23] solution. This observed discrepancy is due to the fact that the surface roughness of the flume bed in the Ritter [23] solution is ignored (fully smooth bed). For the case of the dry bed conditions and at the beginning of a dam-break wave surge, roughness plays a significant role as there is direct contact between the bore front and flume surface. The experimental results agree with a study conducted by Lauber and Hager [42] where the bed roughness was shown to have a significant effect close to the wave fronts. They further concluded what other studies, i.e., Wüthrich et al. [26] also found, namely Ritter's solution does not accurately represent the dam-break flow at the vicinity of the wave front because of the bed roughness.

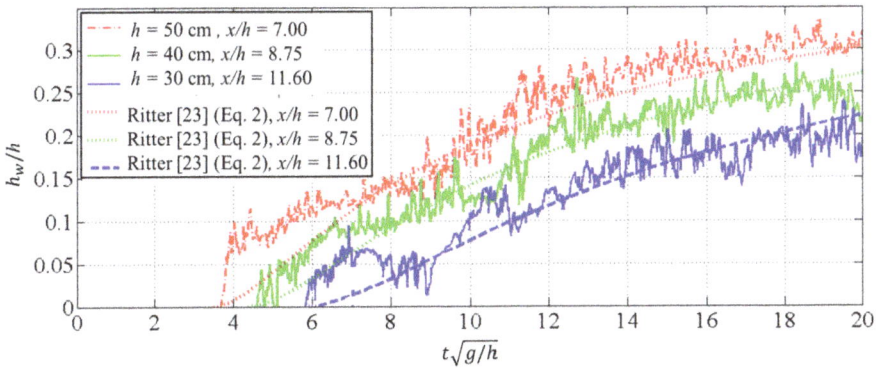

Figure 6. Non-dimensional dry bed condition water time-history surface profile: Comparison with Ritter's (1892) solution. $\frac{h_w}{h}$ versus non-dimensional time $t\sqrt{g/h}$ with $\frac{x}{h} = 7, 8.75, 11.6$.

3.1.2. Dry Bed Bore Front Celerity

The average front celerity of the propagating wave was estimated using the following ratio:

$$U = \frac{\Delta x}{\Delta t}, \tag{3}$$

where U is the bore front celerity in m/s. Δx is the distance between two wave gauges downstream of the gate, WG2 and WG3, and is equal to 2.1 m. The surge travel time is shown with Δt, as the time taken by the bore to travel between the two wave gauge positions.

Several previous studies have estimated the bore front celerity (U) relative to impoundment depth (h) using:

$$U = \alpha\sqrt{gh}, \tag{4}$$

where α is a constant with various values reported in the literature. The constant value depends on the flume hydraulic radius and roughness coefficient. Wüthrich et al. [26] suggest $\alpha = 1.25$, while Matsutomi and Okatamo [4] suggest $\alpha = 1.1$. Figure 7 shows results from the current study, together with the previous studies mentioned above. This study suggests $\alpha = 1.2$ as the constant in Equation (3).

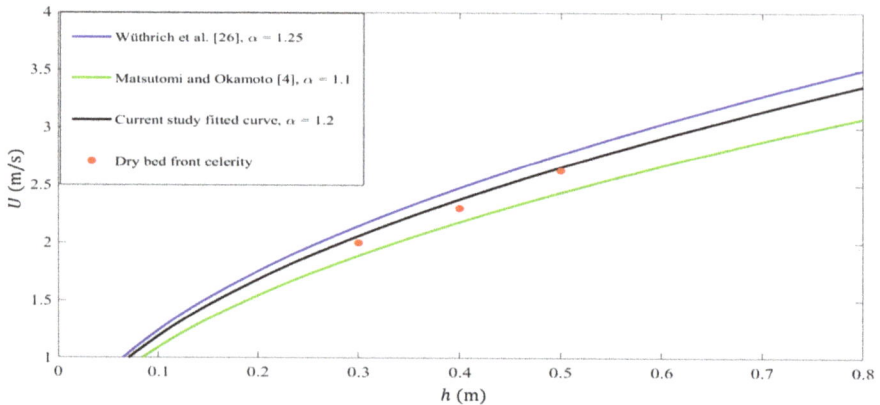

Figure 7. Comparison between the front celerity for dry bed in this study and previous studies.

3.1.3. Dry Bed Flow Velocity, Froude Number and Momentum Flux

Flow characteristics at the pipe location were studied to analyze the exerted forces on the pipeline. Figure 8 illustrates the dry bed surge characteristics as water surface profile (WG3), flow velocity, Froude number and momentum flux at the location of the pipe, respectively.

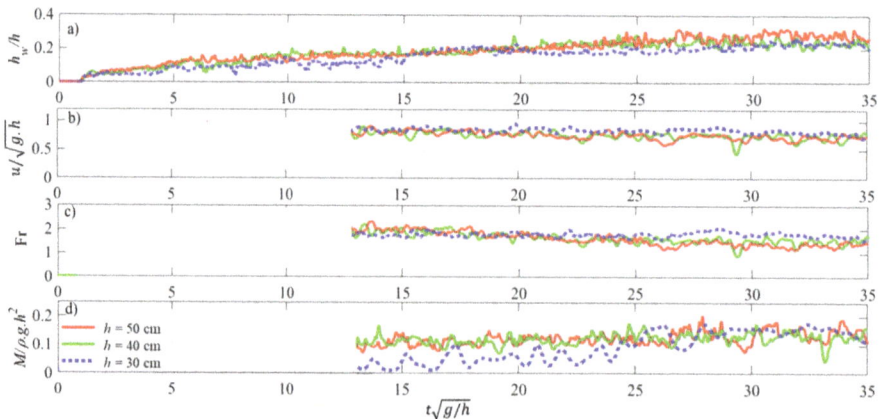

Figure 8. Dry bed surge characteristics at $x = 5.5$ for $h = 30, 40, 50$ cm. (**a**) water surface time-history profile (WG3), (**b**) flow velocity time-history, (**c**) Froude number time-history and (**d**) momentum flux time-history.

The reference time for all the cases is the wave arrival time at WG3. There is a delay in the velocity measurements using the ADV, due to the air entrainment close to the wave arrival time which corresponds to the zone with no data in the first few seconds in Figure 8b–d. Calculated Froude numbers (Fr $= \frac{u}{\sqrt{gh_w}}$) for dry bed conditions and for different reservoir impoundment depths are shown in Figure 8c. It should be noted that for the impoundment depth of $h = 30$ cm, the water level increases and the flow velocity decreases more gradually compared to the cases with $h = 40$ cm and $h = 50$ cm. The dry bed surge was supercritical, Fr > 1, throughout the studied time frame for all three impoundment depths, as shown in Figure 8c. The Froude number remains almost constant in the case of $h = 30$ cm whereas, it gradually decreases for $h = 40$ cm and $h = 50$ cm.

The momentum flux per unit width M is an important factor directly affecting the hydrodynamic loading on structures.

$$M = \rho h_w u^2, \tag{5}$$

where h_w, is the water level and u is the depth-averaged flow velocity. Figure 8d shows the non-dimensional computed momentum flux as $\frac{M}{\rho g h^2}$ for dry bed and different impoundment depths. Figure 8d illustrates that the non-dimensional momentum is smaller for $h = 30$ cm at the beginning of the surge. This could be explained by considerately smaller flow velocity, as well as small water depth in the case of $h = 30$ cm. Smaller momentum flux results in smaller induced drag force as is discussed in the companion paper.

3.2. Wet Bed Condition Hydrodynamics

3.2.1. Wet Bed Water Surface Profile

Figure 9 illustrates the water surface profile for wet bed condition with impoundment depth of $h = 40$ cm and different still water levels (d) at the location of three wave gauges WG1 (a), WG2 (b) and WG3 (c). The reference time in the figure was the gate opening time. The figure shows earlier arrival time for the cases with smaller still water depth (wet bed condition) which indicates a larger bore front celerity in such cases. The values recorded by WG2 and WG3, shown in Figure 9b,c, did not exhibit any change in the water surface profiles. Stoker [43] called the region where the water level remains constant "zone of constant state".

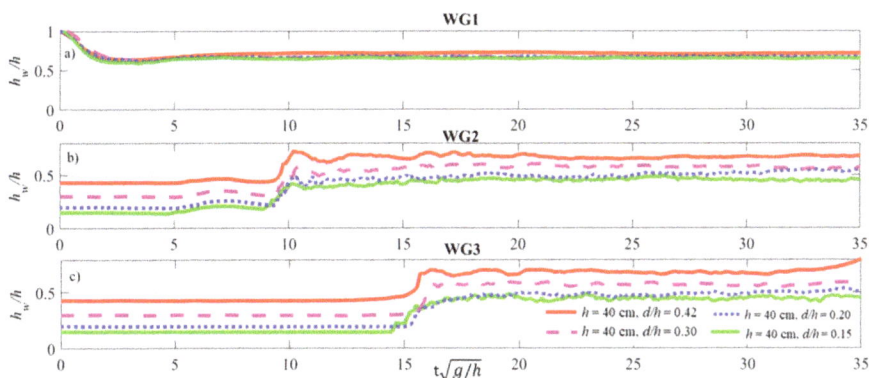

Figure 9. Wet bed bore water surface time history for $h = 40$ cm, different still water levels, $d = 6$, 8, 12, 17 cm, measured at location of (**a**) reservoir wave gauge (WG1), (**b**) $x = 3.5$ m (WG2) and (**c**) $x = 5.5$ m (WG3).

Figure 10 shows a comparison between dry bed and wet bed condition water surface time-histories. The data shows a steeper bore front and more abrupt water level rise in the case of the bore propagating over wet bed when compared to dry bed. Other researchers, i.e., Nouri et al. [29] and Wüthrich et al. [26], also found a similar behavior. According to Wüthrich et al. [26] the behavior of a bore propagating on the wet bed at the wave front, is similar to a turbulent and highly aerated hydraulic jump which causes a more abrupt water level rise compared to wave propagating over dry bed condition.

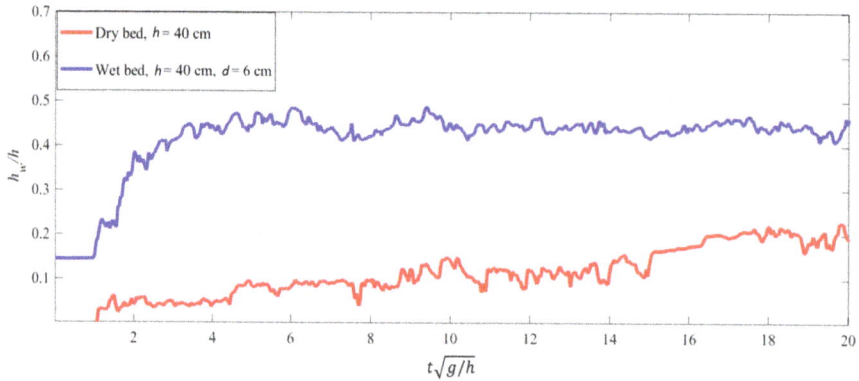

Figure 10. Comparison between dry bed and wet bed condition normalized water depth time-history.

3.2.2. Wet Bed Bore Front Celerity

The bore front celerity was calculated using Equation (3) for the case of wet bed condition for different reservoir impoundment and downstream still water depths. Figure 11 illustrates the dimensionless bore front celerity versus head ratio *(d/h)* obtained from this study together with Chanson [25] empirical solution for bore front celerity in a horizontal channel initially filled with water as:

$$\frac{U}{\sqrt{gd}} = \frac{0.6345 + 0.3286(\frac{d}{h})^{0.65167}}{0.00251 + (\frac{d}{h})^{0.65167}}. \tag{6}$$

Results show good agreement between experimental data and the solution of Chanson [25], which depicts the validity of the empirical solution proposed by this author.

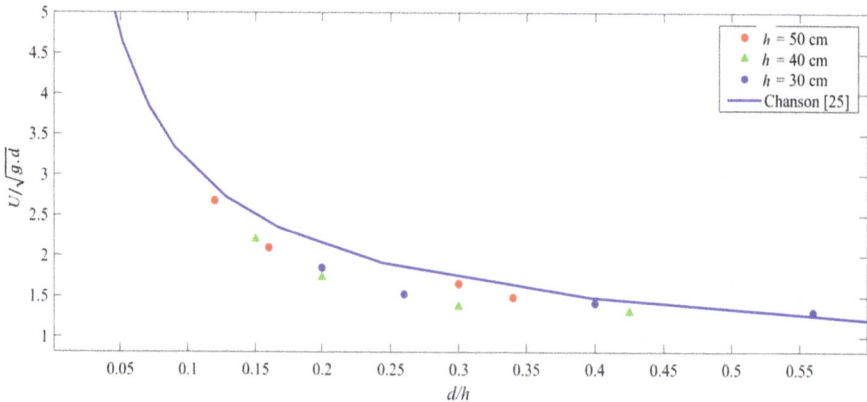

Figure 11. Bore front celerity *d/h*. The solid line shows Chanson [25] solution (Equation (6)), while the points show experimental data.

3.2.3. Wet Bed Flow Velocity, Froude Number and Momentum Flux

Figure 12 depicts the wet bed bore flow velocity, the computed Froude numbers and momentum flux in the case of wet bed condition, a constant impoundment depth of *h* = 40 cm, and different still water depths (*d*). The reference time for all the cases is the wave arrival time at WG3. Due to the air entrainment close to the wave arrival time, the flow velocity data at the beginning of the bore propagation were considered invalid and were eliminated from Figure 12b–d. Results show a noticeable decrease in flow velocity (Figure 12b) and the estimated Froude number (Figure 12c), with an increase in the still water depth or d/h ratio.

This is because such waves were generated using a smaller pressure head (the small difference between the upstream impoundment depth and the downstream still water depth) which resulted in slower flow velocities and lower Froude numbers. Results from all the three tested impoundment depths, i.e., $h = 30, 40, 50$ cm, show that for d/h 0.3 the flow was subcritical, while for $d/h \leq 0.2$ the flow was supercritical. Head ratio values around 0.2 resulted in critical flow regime with Froude number values fluctuating around one. Figure 12d shows the calculated non-dimensional momentum flux $\left(\frac{M}{\rho g h^2}\right)$ values in the case of wet bed condition for impoundment depth of $h = 40$ cm.

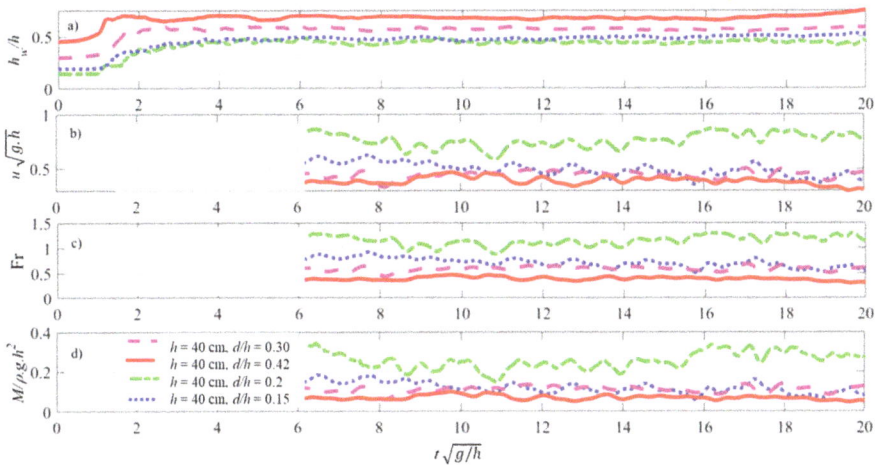

Figure 12. Wet bed bore characteristics at $x = 6.5$ for $h = 40$ cm, different line types show different d/h values. (**a**) Water surface time-history profile (WG3), (**b**) flow velocity time-history, (**c**) Froude number time-history and (**d**) momentum flux time-history.

Results show that the momentum flux decreases as the value of d/h increases. Smaller flow velocities in the bores generated using a smaller head (larger d/h) resulted in smaller momentum values. The same trend was observed for the case with $h = 30$ cm and $h = 50$ cm.

Chanson [25] presented a solution for a dam-break wave moving over a frictionless horizontal channel initially filled with water. The basic flow equations in wet bed conditions are the characteristic system of equations for simple waves as forward and backward characteristics. The forward characteristic in wet bed condition satisfies:

$$V_2 + 2\sqrt{gh_w} = V_0 + 2\sqrt{gh}, \tag{7}$$

where h is the reservoir depth and V_2 and h_2 are the flow velocity (m/s) and bore depth (m) immediately behind the positive surge. The quantity V_0 is the initial reservoir velocity (m/s) equal to zero in the current experiments. Chanson [25] solved this equation together with the continuity and momentum equations graphically. Figure 13 compares Chanson [25] graphical solution and V_2 values measured in current experimental work in a test with $d/h = 0.2$ ($h = 40$ cm, $d = 8$ cm). Results showed good agreement between the two studies.

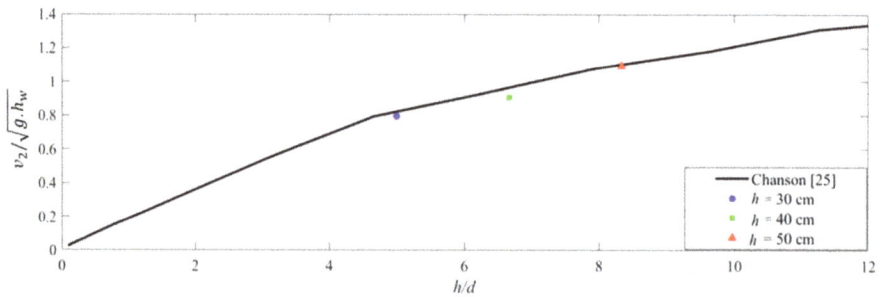

Figure 13. Tests with the wet bed condition flow velocity compared to Chanson's [25] graphical solution.

3.3. Changes in Hydrodynamic Conditions Due to the Presence of the Pipe

3.3.1. Dry Bed Condition

Influence of Pipe Gap Ratio (*e/D*) in Dry Bed Condition

Experimental results in the presence of the pipe showed a considerable change in flow hydrodynamic characteristics, i.e., in the water level and flow velocity. Figure 14 illustrates the alterations in the flow hydrodynamics in dry bed condition in the presence of the pipe with three different gap ratios (*e/D*) compared to the flow in the absence of the pipe.

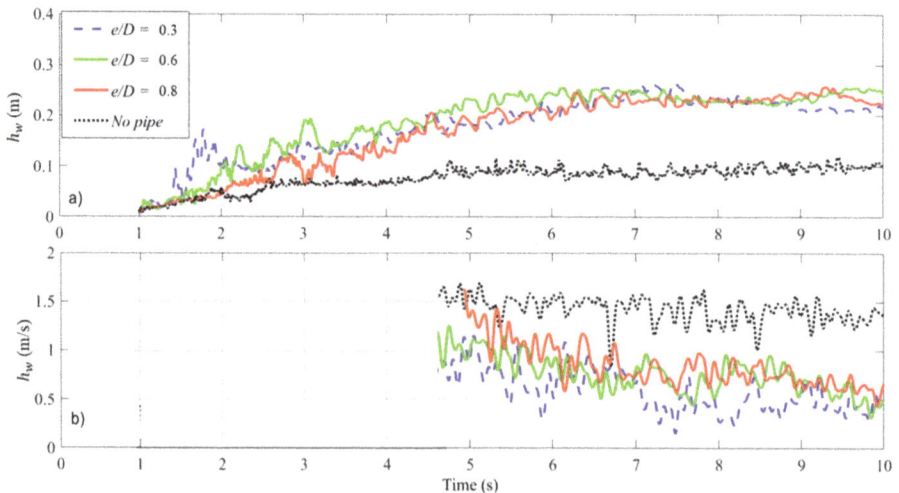

Figure 14. Effect of pipe existence in flow hydrodynamics for different *e/D* values, dry bed condition and *h* = 40 cm. (**a**) Water level time history and (**b**) flow velocity time-history.

The figure also shows that in the case of the smallest gap ratio (*e/D* = 0.3), water reached to the pipe surface faster than other cases and caused abrupt water level rise, as shown in Figure 14 (dashed line at *t* = 1.8 s). The water level rise at the pipe location was due to the flow being blocked by the pipe at the time of wave impact. Results of the flow velocities presented in Figure 14b show that the flow velocity reduced in the presence of the pipe, due to flow blockage by the latter. The flow velocity decreased to smaller values after the partial blockage by the pipe. This happened faster as the gap ratio *e/D* decreased from 0.8 to 0.3, since the water reached the lower edge of the pipe and was blocking the incoming flow sooner. At lower gap ratios, i.e., *e/D* = 0.3, the pipe also got fully submerged earlier which reduced the flow turbulence caused by wave run-up and resulted in the flow velocities to be lower compared to the cases with larger e/D (Figure 14b).

Influence of Impoundment Depth

Figure 15 illustrates the water level rise at the time of bore impacting the pipe. The water level rise is more pronounced for the bore produced by the $h = 50$ cm impoundment depth (Figure 15a) and smallest for that generated by the $h = 30$ cm impoundment depth (Figure 15c). Such abrupt water level rise in larger impoundment depth results in more pronounced impulse force in such cases. Force time-histories are discussed in more detail in the companion paper [41].

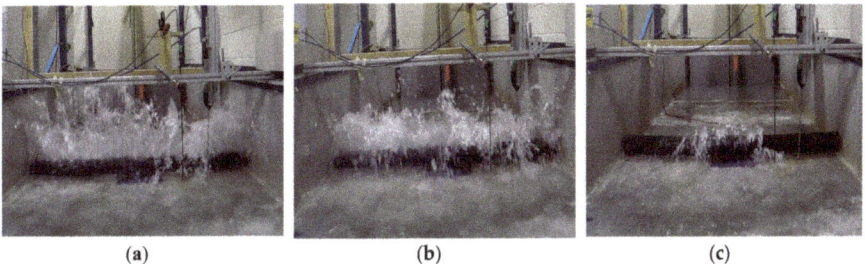

(a) (b) (c)

Figure 15. Water level rise at the time of bore impact generated by impoundment depths of (**a**) $h = 50$ cm, (**b**) $h = 40$ cm and (**c**) $h = 30$ cm.

3.3.2. Wet Bed Condition

Influences of Changing Still Water Depth (d) and Submergence Ratio (S/D)

Figure 16 shows the change in flow hydrodynamics in the case of wet bed condition for different still water depths and submergence ratios (S/D). Results for all different d/h values showed a similar behaviour as in the dry bed case where the water level increased, and flow velocity decreased by the obstruction caused by the presence of the pipe. Figure 16 also shows that in the presence of pipe, the difference between water level and flow velocity with and without pipe is larger in larger still water depths. The root mean square errors (RMSEs) were calculated for flow velocity values and for different still water depths. The calculated RMSE showed a decrease from 0.56 for $d/h = 0.15$, where the pipe is non-submerged ($S/D = 0$, Figure 16e), to 0.19 for $d/h = 0.425$, where the pipe is fully submerged ($S/D = 1$, Figure 16h). RMSE also decreased considerably from 0.40 for the case of less than half submerged pipe ($S/D = 0.2$, Figure 16f) to 0.25 for the case of more than half submerged ($S/D = 0.6$, Figure 16g). Decreased level of pipe submergence resulted in decreased pipe's effective contact area; hence, reduced flow blockage by the pipe. Therefore, the increased level of pipe submergence by increasing still water depth resulted in the reduced influence of pipe presence on flow hydrodynamics. Figure 17 illustrates how the water level rise becomes more abrupt, distinguished with more splashes, at the time of bore impact for cases with lower initial still water depth; thus, smaller level of pipe submergence.

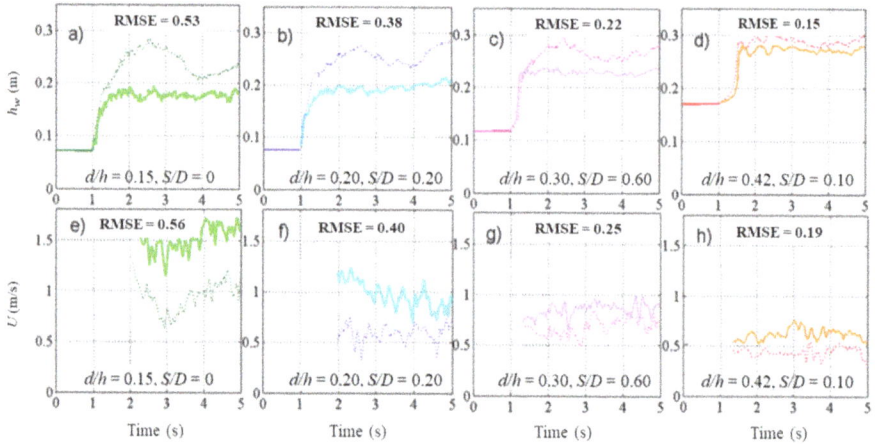

Figure 16. Effect of pipe presence on flow hydrodynamics, different still water depth (d), wet bed condition and impoundment depth, $h = 40$ cm, (**a**) and (**d**): $d/h = 0.15$, non-submerged pipe $S/D = 0$, (**b**) and (**f**): $d/h = 0.2$, less than half submerged $S/D = 0.2$, (**c**) and (**g**): $d/h = 0.3$, more than half submerged $S/D = 0.6$, (**d**) and (**h**): $d/h = 0.425$, fully submerged $S/D = 1.0$. (Dashed line and continuous line show the "with pipe" and "no pipe" conditions, respectively).

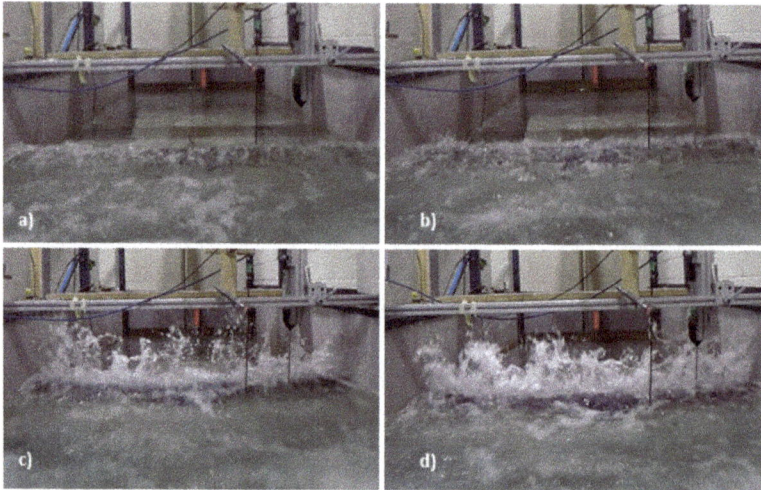

Figure 17. Water level rise at the time of bore impact, wet bed condition, and $h = 40$ impoundment depth cm (**a**) $d/h = 0.425$, $S/D = 1$, (**b**) $d/h = 0.3$, $S/D = 0.6$, (**c**) $d/h = 0.2$, $S/D = 0.2$, (**d**) $d/h = 0.2$, $S/D = 0$.

3.4. Scale Effects

Table 3 summarizes the flow conditions in this study at the location of the pipe. Previous modelling results from the 1993 Hokkaido-Nansei-Oki Tsunami estimated water depths in the range of 5–15 m and flow velocities in the range of 3–15 m/s. As shown in Table 3, the wave height and wave front celerity measured in this study are in the range of the modelling results mentioned above. According to Lauber and Hager [42], for impoundment depths $h \geq 0.30$ m, inertia forces are dominating and the flow is governed by Froude similarity. This condition applies to all experiments conducted in this study, where the impoundment depths were either equal to or larger than 30 cm. According to Bricker et al. [44], when using Froude scaling for tsunami modelling, surface tension and viscous effects must be appropriately considered. Weber numbers in Table 3 were calculated using:

$$We = \frac{\rho u^2 h_w}{\sigma} \tag{8}$$

where ρ is the water density ($\frac{kg}{m^3}$), u is the flow velocity, h_w is the water depth, and σ is the surface tension (N/m).

Table 3. Hydrodynamic conditions at the pipe location.

Reservoir Depth h (m)	Head Ratio d/h	Maximum Wave Height (m)	Wave Front Celerity (m/s)	Weber Number (-)	Flow Reynolds Number (-)	Pipe Reynolds Number (-)
	0.00	0.078	2.00	4285	1.56×10^5	2.00×10^5
	0.200	0.100	1.41	2730	1.41×10^5	1.41×10^5
0.3	0.260	0.107	1.53	3440	1.56×10^5	1.53×10^5
	0.400	0.136	1.35	3404	1.83×10^5	1.35×10^5
	0.560	0.163	1.67	6244	2.7×10^5	1.67×10^5
	0.000	0.128	2.27	9060	2.90×10^5	2.26×10^5
	0.150	0.172	1.68	6688	2.88×10^5	1.68×10^5
0.4	0.200	0.176	1.53	5660	2.69×10^5	1.53×10^5
	0.300	0.224	1.48	6739	3.31×10^5	1.48×10^5
	0.425	0.268	1.67	10,266	4.47×10^5	1.67×10^5
	0.000	0.160	2.60	14,857	4.16×10^5	2.60×10^5
	0.120	0.215	2.05	12,411	4.40×10^5	2.05×10^5
0.5	0.160	0.220	1.85	10,342	4.07×10^5	1.85×10^5
	0.240	0.280	1.79	12,323	5.01×10^5	1.79×10^5
	0.340	0.335	1.91	16,787	6.39×10^5	1.91×10^5

As shown in Table 3, We in all the tested cases are larger than the critical We defined by Peakall and Warburton [45], i.e., $We_{crit} \leq 120$. Therefore, the effect of changing surface tension in relation to nature scale could be neglected in this study. According to Te Chow [46], the flow in this study is fully turbulent ($1.41 \times 10^5 < Re < 6.39 \times 10^5$). However, tsunami flow is usually associated with flow Reynolds numbers larger than 10^6. Therefore, the bottom boundary layer may not be properly represented in the conducted experiments [44]. The present study focuses on the force on the pipe; thus, pipe Re could be more influential. The flow and pipe Reynolds number in the experiments were calculated using:

$$Re = \frac{h_w u}{\nu}, \tag{9}$$

$$Re = \frac{D u}{\nu}, \tag{10}$$

where h_w is the water depth, D is the pipe diameter and ν is the kinematic viscosity. The flow velocity (u) in Equation (8) was the free stream velocity measured at the location of the pipe center. Calculated values were in the range of $8 \times 10^4 < Re < 2.6 \times 10^5$. Wüthrich et al. [35] and Sumer and Fredsøe [47] characterized the flow around the cylinder in this range of Re with a completely turbulent wake and a laminar boundary layer separation both of which cause high pressure and large pressure drag in front of the cylinder. Therefore, turbulent wake flow, which plays an important role on the induced forces on the pipe, is well represented in the experiments.

4. Conclusions

The results of this study constitute the first part of a two-part work, presenting an experimental study on the impact of dam-break tsunami-induced hydraulic bores interacting with horizontally-mounted pipelines. The focus of this first part was on the flow hydrodynamics in dry and wet bed conditions and its changing characteristics in the presence of the pipe in the flow. The following conclusions are drawn from this research:

- For the dry bed condition, the bore front celerity increased with an increase in the impoundment depth. $\alpha = 1.2$ was suggested to be used in Equation (4) for the bore front celerity expression.
- The water surface profile and flow velocity, as well as the flow Froude number, were shown to change more gradually over the same period of time for small impoundment depths (i.e., $h = 30$ cm) compared to the waves generated by higher impoundment depths. Momentum flux was also smaller in the wave front region for $h = 30$ cm, due to a smaller flow velocity and water depth.
- Increasing the still water level downstream of the gate led to slower bore flow velocities, reduced Froude number, and reduced momentum flux compared to the bore produced by the same impoundment depth, but propagating over the dry bed. The flow regime changes from supercritical to subcritical with an increase in the still water depth and for $d/h > 0.3$.
- The presence of the pipe, for both dry and wet bed condition, caused the water level to rise and the flow velocity to decrease. In dry bed condition, smaller e/D values resulted in more abrupt water level rise at the time of the bore impact and a faster decrease in flow velocity.
- For bore propagating over dry bed, the water level increase at the time of bore impact in the presence of the pipe became larger with an increasing impoundment depth.
- In the case of the wet bed condition, increased level of pipe submergence S/D, due to increasing the still water depth d resulted in a reduction of the influence of the pipe on flow hydrodynamics. This was explained by a reduction in the flow blockage, due to the increased pipe submergence.

Author Contributions: B.G. developed the methodology and carried out the experiments. B.G. was also the main responsible for the analysis of the data and writing the manuscript. J.S. assisted in conducting the experiments and contributed in reviewing and editing. I.N. and A.M. conceived of the presented idea, supervised the work and contributed in reviewing and editing. N.G. provided some of the instruments utilized in this study, assisted in the experiments and also contributed in reviewing and editing.

Funding: This research was funded by NSERC Discovery Grants held by Ioan Nistor, No. 210282 and Majid Mohammadian, No. 210717. Partial support for the study came through the Marie Curie International Outgoing Fellowship of Nils Goseberg within the 7th European Community Framework Program, No. 622214).

Acknowledgments: The authors are grateful to the University of Ottawa Hydraulic Laboratory Technician, Mark Lapointe, as well as to Adrian Simpalean and Derek Eden, graduate students at the University of Ottawa, for their assistance during the experimental work.

Conflicts of Interest: The authors declare no conflict of interest.

References

1. Nicholls, R.J. Coastal flooding and wetland loss in the 21st century: Changes under the SRES climate and socio-economic scenarios. *Glob. Environ. Change* **2004**, *1*, 69–86. [CrossRef]
2. Cao, Z.; Yue, Z.; Pender, G. Landslide dam failure and flood hydraulics. Part I: Experimental investigation. *Nat. Hazards* **2011**, *2*, 1003–1019. [CrossRef]
3. Prasad, S. Wave Impact Forces on a Horizontal Cylinder. Ph.D. Dissertation, University of British, Columbia, UK, 1994.
4. Matsutomi, H.; Okamoto, K. Inundation flow velocity of tsunami on land. *Island Arc.* **2010**, *3*, 443–457. [CrossRef]
5. Dias, P.; Dissanayake, R.; Chandratilake, R. Lessons Learned from Tsunami Damage in Sri Lanka. *Civil Eng.* **2006**, *159*, 74–81. [CrossRef]
6. Rossetto, T.; Peiris, N.; Pomonis, A.; Wilkinson, S.M.; Re, D.; Koo, R.; Gallocher, S. The Indian Ocean tsunami of December 26, 2004: Observations in Sri Lanka and Thailand. *Nat. Hazards.* **2007**, *1*, 105–124. [CrossRef]

7. Jaffe, B.E.; Goto, K.; Sugawara, D.; Richmond, B.M.; Fujino, S.; Nishimura, Y. Flow speed estimated by inverse modeling of sandy tsunami deposits: Results from the 11 March 2011 tsunami on the coastal plain near the Sendai Airport, Honshu, Japan. *Sediment. Geol.* **2012**, *282*, 90–109. [CrossRef]
8. Fritz, H.M.; Phillips, D.A.; Okayasu, A.; Shimozono, T.; Liu, H.; Mohammed, F.; Skanavis, V.; Synolakis, C.E.; Takahashi, T. The 2011 Japan tsunami current velocity measurements from survivor videos at Kesennuma Bay using LiDAR. *Geophys. Res. Lett.* **2012**, *39*. [CrossRef]
9. Limura, K.; Norio, T. Numerical simulation estimating effects of tree density distribution in coastal forest on tsunami mitigation. *Ocean Eng.* **2012**, *54*, 223–232. [CrossRef]
10. Synolakis, C.E. The run-up of solitary waves. *J. Fluid Mech.* **1987**, *185*, 523–545. [CrossRef]
11. Gedik, N.; Irtem, E.; Kabdasli, S. Laboratory investigation on tsunami run-up. *Ocean Eng.* **2005**, *32*, 513–528. [CrossRef]
12. Hsiao, S.C.; Lin, T.C. Tsunami-like solitary waves impinging and overtopping an impermeable seawall: Experiment and RANS modeling. *Coast. Eng.* **2010**, *57*, 1–18. [CrossRef]
13. Aristodeme, F.; Tripepi, G.; Meringolo, D.D.; Veltri, P. Solitary wave-induced forces on horizontal circular cylinders: Laboratory experiments and SPH simulations. *Coast. Eng.* **2017**, *129*, 17–35. [CrossRef]
14. Madsen, A.; Schäffer, H.A. Analytical solutions for tsunami run-up on a plane beach: Single waves, N-waves and transient waves. *J. Fluid Mech.* **2010**, *645*, 27–57. [CrossRef]
15. Madsen, A.; Fuhrman, D.R.; Schäffer, H.A. On the solitary wave paradigm for tsunamis. *J. Geophys. Res.* **2008**, *113*, 286–292. [CrossRef]
16. Chan, I.C.; Liu, L.F. On the run-up of long waves on a plane beach. *J. Geophys. Res.* **2012**, *117*, 72–82. [CrossRef]
17. Leschka, S.; Oumeraci, H. Solitary waves and bores passing three cylinders-effect of distance and arrangement. *Coast. Eng. Proc.* **2014**, *1*, 23. [CrossRef]
18. Istrati, D.; Buckle, I.; Lomonaco, P.; Yim, S. Deciphering the tsunami wave impact and associated connection forces in open-girder coastal bridges. *J. Mar. Sci. Eng.* **2018**, *6*, 148. [CrossRef]
19. Zhao, E.; Qu, K.; Mu, L.; Kraatz, S.; Shi, B. Numerical study on the hydrodynamic characteristics of submarine pipelines under the impact of real-world tsunami-like waves. *Water* **2019**, *11*, 221. [CrossRef]
20. Chanson, H.; Aoki, S.I.; Maruyama, M. An experimental study of tsunami runup on dry and wet horizontal coastlines. *Sci. Tsunami Hazards* **2003**, *20*, 278–293.
21. Chanson, H. Tsunami surges on dry coastal plains: Application of dam break wave equations. *Coast. Eng. J.* **2006**, *48*, 355–370. [CrossRef]
22. Stolle, J.; Ghodoosipour, B.; Derschum, C.; Nistor, I.; Petriu, E.; Goseberg, N. Swing gate generated dam-break waves. *J. Hydraul. Res.* **2018**, 1–13. [CrossRef]
23. Ritter, A. Die Fortpflanzung der Wasserwellen. *Z Des. Vereines Dtsch. Ingenieure* **1892**, *36*, 947–954.
24. Henderson, F.M. *Open Channel Flow*; MacMillan Company: New York, NY, USA, 1966.
25. Chanson, H. Applications of the Saint-Venant equations and method of characteristics to the dam break wave problem. *Hydraul. Model Rep.* **2005**, *47*, 41–49. [CrossRef]
26. Wüthrich, D.; Pfister, M.; Nistor, I.; Schleiss, A.J. Experimental study of tsunami-like waves generated with a vertical release technique on dry and wet beds. *J. Water W Port Coast* **2018**, *4*, 04018006. [CrossRef]
27. St-Germain, P.; Nistor, I.; Townsend, R.; Shibayama, T. Smoothed-particle hydrodynamics numerical modelling of structures impacted by tsunami bores. *J. Water W Port Coast* **2013**, *1*, 66–81.
28. Douglas, S.; Nistor, I. On the effect of bed condition on the development of tsunami-induced loading on structures using OpenFOAM. *Nat. Hazards* **2015**, *2*, 1335–1356. [CrossRef]
29. Nouri, Y.; Nistor, I.; Palermo, D. Experimental investigation of tsunami impact on free standing structures. *Coast. Eng. J.* **2010**, *52*, 43–70. [CrossRef]
30. Al-Faesly, T.; Palermo, D.; Nistor, I.; Cornett, A. Experimental modeling of extreme hydrodynamic forces on structural models. *Int. J. Pro. Str.* **2012**, *3*, 477–506. [CrossRef]
31. Bremm, G.C.; Goseberg, N.; Schlurmann, T.; Nistor, I. Long wave flow interaction with a single square structure on a sloping beach. *J. Mar. Sci. Eng.* **2015**, *3*, 821–844. [CrossRef]
32. Foster, A.S.J.; Rossetto, T.; Allsop, W. An experimentally validated approach for evaluating tsunami inundation forces on rectangular buildings. *Coast. Eng. J.* **2017**, *128*, 44–57. [CrossRef]
33. Wüthrich, D.; Pfister, M.; Nistor, I.; Schleiss, A.J. Experimental study on the hydrodynamic impact of tsunami-like waves against impervious free-standing buildings. *Coast. Eng. J.* **2018**, *60*, 180–199. [CrossRef]

34. Wüthrich, D.; Pfister, M.; Nistor, I.; Schleiss, A.J. Experimental study on forces exerted on buildings with openings due to extreme hydrodynamic events. *Coast. Eng. J.* **2018**, *140*, 72–86. [CrossRef]

35. Arnason, H. Interactions Between an Incident Bore and a Free-Standing Coastal Structure. Ph.D. Dissertation, University of Washington, Seattle, WA, USA, 2005.

36. Goseberg, N.; Schlurmann, T. Non-stationary flow around buildings during run-up of tsunami waves on a plain beach. In Proceedings of the Coastal Engineering Conference, American Society of Civil Engineers (ASCE), Reston, VA, USA, 15–20 June 2014.

37. Araki, S.; Deguchi, I. Characteristics of wave pressure and fluid force acting on bridge beam by tsunami. *Coast. Struct.* **2011**, 1299–1310. [CrossRef]

38. Mazinani, I.; Ismail, Z.; Hashim, A.M.; Saba, A. Experimental investigation on tsunami acting on bridges. *Int. J. Civ. Env. Eng.* **2014**, *8*, 1040–1043.

39. Chen, C.; Melville, B.W.; Nandasena, N.A.K.; Farvizi, F. An experimental investigation of tsunami bore impacts on a coastal bridge model with different contraction ratios. *J. Coast. Res.* **2017**, *34*, 460–469. [CrossRef]

40. Louis, S.T. *Missouri Minimum Design Loads and Associated Criteria for Buildings and Other Structures*; ASCE/SEI (ASCE/Structural Engineering Institute): Reston, VA, USA, 2017; pp. 25–50.

41. Ghodoosipour, B.; Stolle, J.; Nistor, I.; Mohammadian, A.; Goseberg, N. Experimental study on extreme hydrodynamic loading on pipelines part 2: Induced force analysis. *JMSE* **2019**.

42. Lauber, G.; Hager, W.H. Experiments to dam break wave: Horizontal channel. *J. Hydraul. Res.* **1998**, *3*, 291–307. [CrossRef]

43. Stoker, J.J. *Water Waves: The Mathematical Theory with Applications*; John Wiley Sons: Hoboken, NJ, USA, 1958.

44. Bricker, J.D. On the need for larger Manning's roughness coefficients in depth-integrated tsunami inundation models. *Coast. Eng. J.* **2015**, *57*, 1550005. [CrossRef]

45. Peakall, J.; Warburton, J. Surface tension in small hydraulic river models- The significance of the Weber number. *J. Hydrol. (New Zealand)* **1998**, *35*, 199–212.

46. Chow, V. *Open Channel Hydraulics*; McGraw-Hill Book Company, Inc.: New York, NY, USA, 1959.

47. Sumer, B.M.; Fredsøe, J. *Hydrodynamics Around Cylindrical Structures*; World Scientific Publishing Company: Singapore, 2006.

© 2019 by the authors. Licensee MDPI, Basel, Switzerland. This article is an open access article distributed under the terms and conditions of the Creative Commons Attribution (CC BY) license (http://creativecommons.org/licenses/by/4.0/).

Journal of
Marine Science and Engineering

MDPI

Article

Experimental Study on Extreme Hydrodynamic Loading on Pipelines Part 2: Induced Force Analysis

Behnaz Ghodoosipour [1,*], Jacob Stolle [1], Ioan Nistor [1], Abdolmajid Mohammadian [1] and Nils Goseberg [2]

[1]   Department of Civil Engineering, University of Ottawa, Ottawa, ON K1N 6N5 Canada
[2]   Leichtweiß-Institute for Hydraulic Engineering and Water Resources, Technische Universität Braunschweig, 38124 Braunschweig, Germany
*   Correspondence: bghod068@uottawa.ca; Tel.: +1-613-562-5800 (ext. 6159)

Received: 23 May 2019; Accepted: 6 August 2019; Published: 9 August 2019

Abstract: Adequate design of pipelines used for oil, gas, water, and wastewater transmission is essential not only for their proper operation but particularly to avoid failure and the possible extreme consequences. This is even more drastic in nearshore environments, where pipelines are potentially exposed to extreme hydrodynamic events, such as tsunami- or storm-surge-induced inundation. The American Society of Civil Engineers (ASCE), in its ASCE7 Chapter 6 on Tsunami Loads and Effects which is the new standard for tsunami impacts and loading, specifically stresses the need to study loads on pipelines located in tsunami-prone areas. To address this issue, this study is the first of its kind to investigate loading on pipelines due to tsunami-like bores. A comprehensive program of physical model experiments was conducted in the Dam-Break Hydraulic Flume at the University of Ottawa, Canada. The tests simulated on-land tsunami flow inundation propagating over a coastal plain. This allowed to record and investigate the hydrodynamic forces exerted on the pipe due to the tsunami-like, dam-break waves. Different pipe configurations, as well as various flow conditions, were tested to investigate their influence on exerted forces and moments. The goal of this study was to propose, based on the results of this study, resistance and lift coefficients which could be used for the design of pipelines located in tsunami-prone areas. The values of the resistance and lift coefficients investigated were found to be in the range of $1 < C_R < 3.5$ and $0.5 \leq C_L < 3$, respectively. To that end, the study provides an upper envelope of resistance and lift coefficients over a wide range of Froude numbers for design purposes.

Keywords: pipelines; extreme event; tsunami; dam-break wave; drag force; force coefficient

1. Introduction

Pipelines located in coastal areas are important infrastructures which are used for gas and oil transportation, as well as for conveying and disposing of potable and wastewater, respectively. The Canadian Energy Pipeline Association [1] published a report on "pipeline watercourse management" which focuses on damage to pipelines caused by natural hazards and proposes practices to ensure the safety of public and environment in the case of severe damages to operating pipelines in Canada. Safe design of pipelines in coastal areas is of critical importance, as damage to these pipelines can have significant economic and environmental consequences. Different engineering criteria need to be considered as pipes located in the vicinity of coastal waters or in the shallow water region are subject to a variety of regular loading conditions, such as waves, tides, and nearshore currents, in addition to extreme loads, such as storm surges and tsunamis. Until now, the focus of research on pipe loading exclusively focused on hydrodynamic forces exerted in steady unidirectional or oscillatory flows for horizontal or vertical cylinders.

J. Mar. Sci. Eng. **2019**, *7*, 262

To study the hydrodynamic forces exerted on pipelines, understanding the flow behavior around the pipe is critical. Several studies on flow behavior around circular cylinders near a plane boundary (wall) were performed.

Flow around a circular cylinder changes depending on the flow characteristics defined, among other parameters, by the cylinder Reynolds number.

$$Re = \frac{Du}{v},$$ (1)

where D is the diameter of the cylinder, u is the flow velocity, and v is the kinematic viscosity. The flow undertakes considerable changes with an increase in Re as the wake and boundary layer characteristics experience massive change. Details of such changes were given by Sumer and Fredsøe [2].

Vortex shedding behind an isolated circular cylinder, which occurs when flow turbulence reaches above a critical Reynolds number (Re = 40), is a commonly observed phenomenon which causes uncertainties in the estimation of the drag and lift forces. Due to the frequent vortex shedding occurring behind the pipe placed in flow, local pressure along the cylinder circumference changes and, hence, causes superimposing fluctuations to the drag and lift forces exerted on the cylinder [3]. Drag and lift forces are usually expressed by their non-dimensional forms as drag and lift coefficient, respectively. Drag and lift coefficients depend on both the cylinder geometry and the incoming flow characteristics. In the case of horizontal cylinders close to plane boundaries, the turbulent characteristics of the flow in the region close to that boundary change depending on the boundary properties such as its roughness. Such variations in the flow characteristics affect the abovementioned non-dimensional parameters.

The effect of the wall proximity on the exerted hydrodynamic forces, as well as the properties of the vortex shedding on horizontal circular cylinders, made the subject of intense research in the past. When pipelines are placed on the seabed, flow around the pipe may cause scouring of the bed material. This may lead to the pipe being suspended above the bed with an underlying gap, usually in the range from $0.1D$ to $1.0D$ [2]. Above-ground transmission pipelines, installed in initially dry conditions, are also usually installed onto supports which ensure a prescribed distance between the pipe and the ground. When such pipes are accidentally flooded, this distance can significantly impact the flow around the pipeline. Therefore, it is important to understand the flow characteristics that occur in the proximity of such a pipe.

Several researchers studied the induced forces on pipes placed near seabeds, such as Aristodemo et al. [4] who proposed a new numerical model for the estimation of the horizontal and vertical hydrodynamic forces induced on submarine pipelines exposed to non-linear wave and current conditions.

The parameter gap ratio, e/D (e being the distance between the lower edge of the pipe and the ground), is defined for the purpose of investigating the impact of bed proximity on flow hydrodynamics around the pipe. In the case of large e/D values ($e/D > 1$), the pipe acts similar to a free cylinder with no boundary effects.

Figure 1a defines the location of a stagnation point (a point in a flow field where the local velocity of the fluid is zero) and separation point (the position where a boundary layer separates from the surface of a solid body), while Figure 1b illustrates the displacement of these points in the existence of a plane boundary during a wave and/or current flow around a circular cylinder.

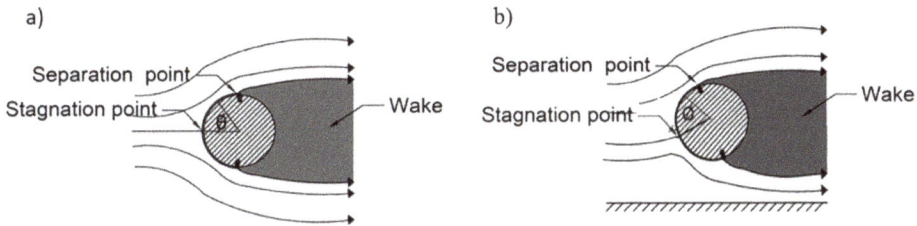

Figure 1. Schematic of the flow around a circular cylinder, showing stagnation and separation points: (**a**) free cylinder, and (**b**) cylinder near a plane boundary (adapted from Reference [3]).

According to Sumer and Fredsøe [2], flow around a cylinder placed close to the bed as shown in Figure 1b may change as follows depending on the gap ratio e/D:

The angular position of the stagnation point $\varnothing$ displaces and moves to lower angles.

A change in the gap ratio also changes the angular position of the separation point. As the gap ratio decreases, the separation point at the free-stream side of the cylinder moves upstream, while the separation point at the wall side moves downstream.

Vortex shedding suppression happens when $e/D < 0.3$ due to the asymmetry of the generated vortices on the free stream and wall side of the cylinder. Larger vortices on the free-stream side interact with small vortices occurring near the wall and cause vortex shedding suppression compared to regular vortex shedding, which results in increased suction on the free side of the cylinder.

Most studies revealed a gap ratio of $e/D = 0.3$ as the distance for which vortex shedding suppression commences. Bearman and Zdravkovich [5], Angrilli et al. [6], and Zdravkovich [7] supported this hypothesis by looking into the power spectra of hot-wire anemometer signals occurring in the wake region behind a cylinder. However, results from Buresti and Lanciotti [8] showed a gap ratio of $e/D = 0.4$ to be the critical gap ratio. The existence of vortex shedding causes the occurrence of a peak in the power spectra which disappears due to the suppression at a specific gap ratio.

A new topic in pipeline design emerged due to the recent extreme events, such as the impact of pipelines by coastal flooding generated by tsunamis and storm surges, which caused massive damage to such infrastructure. Catastrophic events, such as the 2011 Tōhoku Tsunami in Japan and the 2012 Typhoon Haiyan in the Philippines, increased the interest of researchers to re-evaluate existing design and safety standards to consider the effects of such extreme events. The American Society of Civil Engineers (ASCE), through its ASCE7 Tsunami Loads and Effects Committee, developed a new standard for tsunami impacts and loading, ASCE7 Chapter 6 [9]. Amongst the potential effects of such extreme events on infrastructure, the standard specifically emphasized the need to investigate tsunami loads on pipelines located in the flood zone.

Several research articles focused on investigating forces induced by a hydraulic bore on infrastructure experimentally. Wind wave forces on cylinders were the subject of many studies in the context of hydrodynamics [2] and coastal engineering studies [10,11]. Qi et al. [12] investigated the tsunami inundation forces exerted on structures using steady flow data. Several other studies [13–17] evaluated the forces induced from tsunami wave–structure interaction under unsteady conditions. Aristodemo et al. [18] conducted a small-scale experimental study in a wave flume along with a numerical investigation using a smoothed-particle hydrodynamics model. In their study, they investigated the induced loading from a solitary wave on a horizontal circular cylinder.

Various guidelines are available for the design of buildings against tsunami loads and effects. Among them, the Federal Emergency Management Agency (FEMA [19]), in the section related to the design of Structures for Vertical Evacuation from Tsunamis, mentions different forces exerted on a body during a tsunami as follows:

- Hydrostatic and buoyant forces: Hydrostatic forces occur when still or slow-moving water interacts with a fixed body. The vertical component of the hydrostatic force is the buoyancy force. For partially or fully submerged bodies, the buoyancy force is exerted at the centroid of the displaced water volume.

- Hydrodynamic force: Also referred to as the drag force, it is caused by fluid flowing around a structure. The structure's geometry, and its flow characteristics and fluid density influence the magnitude and direction of the hydrodynamic force. The drag force is a combination of the pressure force from moving the mass of fluid, as well as friction force between the flowing fluid and the structure. The hydrodynamic force is exerted at the centroid of the wetted surface and can be calculated as

$$F_d = \frac{1}{2}\rho_s C_d B (h_w u^2)_{max},$$ (2)

where ρ_s is the water density, C_d is the drag coefficient, B is the width of the structure in the plane normal to the direction of flow, h_w is the water surface elevation or flow depth, and u is flow velocity at the location of the structure. C_d depends on both flow characteristics and the structure's geometry and orientation. FEMA [19] suggests using a drag coefficient value $C_d = 2$, while the recent update to the ASCE7 standard [9] suggests values based on the width-to-inundation-depth ratio and the type of structural component. Suggested C_d values in the ASCE7 are in the range of $1.2 < C_d < 2.5$.

- Impulse force: Impulse or impact forces act on a body when the leading edge of the bore reaches the structure rapidly. Several studies focused on measuring and analyzing the impulse force on vertical structures. FEMA [19] guideline refers to different studies such as Arnason [14] and Ramsden [20] and suggests an impulse force equal to 1.5 times the hydrodynamic force in the case of structural wall elements. Consequently, such forces cause damage to the structural element, as well as to the joints, and may cause massive damage on the structure even when the hydrodynamic force is larger than the impulse force [11,21]. The impulse force is usually followed by an increase in force magnitude as the bore flow accumulates in front of the structure causing a "bulb-like" wake. This force is also termed "run-up force" or "transient hydrodynamic force" [21].

- Uplift force: Uplift forces are usually exerted on elevated surfaces that are submerged during tsunami inundation waves. In fact, water passes through a gap between the ground and the elevated surface remarkably fast during a tsunami and this induces uplift onto the bottom surface of the elevated horizontal components. This uplift force adds to the hydrostatic vertical component (the buoyancy force). FEMA [19] suggests a formula for computing the uplift force as

$$F_u = \frac{1}{2}C_u \rho_s A_f u_v^2,$$ (3)

where C_u is the lift coefficient, and ρ_s is the fluid density. FEMA [19] suggests $C_u = 3.0$. A_f is the area of the floor panel, and u_v is the estimated vertical velocity or water rise rate.

The abovementioned force components were previously addressed in studies and guidelines for the design of vertical evacuation structures. However, to the authors' knowledge, no study investigated the loads exerted in the case of tsunami inundation waves impacting horizontal circular cylinders or pipelines, and there is a lack of knowledge regarding the characteristics of such loads on the following:

- Above-ground pipelines, placed at a given distance from the ground using supports.
- Fully submerged pipelines placed onto or near the ocean or seabed.
- Partially submerged pipelines.

Objectives

Based on the above outlined lack of knowledge, this study details the findings in this second portion of a two-part work describing a comprehensive experimental program to investigate extreme hydrodynamic loading on pipes. The first part of this study [22] focused on the description of the experimental program and instrumentation, the hydrodynamics of the extreme flow generated, and the influence of the pipe presence on flow hydrodynamics. This second part details the results of the study with respect to the hydrodynamic forces measured and the calculation of the force coefficients. The specific objectives of this second part of the study are as follows:

- Measuring and analyzing the changes in drag and lift forces exerted on the pipe due to a tsunami-like bore using different pipe elevations or gap ratios e/D for the dry bed condition.
- Studying the effect of still water depth d and of the impoundment head-to-depth ratio d/h, d being the still water depth and h being the reservoir depth, on the various force components caused by tsunami-like dam-break wave on the pipelines in the wet bed condition.
- Investigating the variation of the drag and lift forces exerted on fully and partially submerged pipelines with the different initial level of submergence ratio S/D, in order to examine the influence of the submergence on the force components.
- Studying the characteristics and the magnitude of the forces exerted on pipelines by dam-break waves generated using different reservoir impoundment depth h, for both dry and wet bed conditions.
- Proposing force coefficients for various flow conditions and pipe configurations.

It is anticipated that these results will lead to formulating recommendations for the optimal design of pipelines located in tsunami-prone areas, which will be both environmentally and economically safe during such extreme events. The research will also allow for better design and determine load conditions in regions subjected to potential natural hazards.

2. Experimental Set-Up

A comprehensive experimental program was designed and conducted in the Dam-Break Flume in the Hydraulic Laboratory at the University of Ottawa, Canada. The flume is 30.1 m long, 1.5 m wide, and 0.5 m deep. Various hydrodynamic parameters, as well as the forces exerted on the physical model of the horizontal pipeline (0.1 m in diameter and 1.47 m in length), were measured during the tests. Figure 2 illustrates the schematic view of the dam-break flume and location of instruments and the experimental parameters (impoundment depth and downstream water levels).

Details of the experimental facilities and their characteristics are reported in the companion paper [22]. Table 1 presents an overview of the instruments used in the experiments and their specifications.

Table 1. Instruments used in the experimental program.

Instrument	Manufacturer, Model	Sampling Rate	Accuracy
Wave gauge (WG)	RBR WG-50, capacitance-type	300 Hz	±0.002
Acoustic doppler velocimeter (ADV)	Nortek, Vectrino	200 Hz	±1 mm/s
Dynamometer	Interface-6A68E	300 Hz	±0.04%
High-speed camera (HS)	Flare 2M360-CL	200 Hz	-

Experimental Program

A comprehensive experimental study was conducted to investigate effects of design parameters on extreme hydrodynamic loadings on pipelines. Parameters considered in this study include reservoir depth, h, still water depth, d, ratio of the lower edge of pipe distance to bed, e, to diameter, e/D, and pipe level of submergence, S, to pipe diameter ratio, S/D. An initial set of tests was performed without the pipe installed in the flume to investigate the initial hydrodynamic conditions.

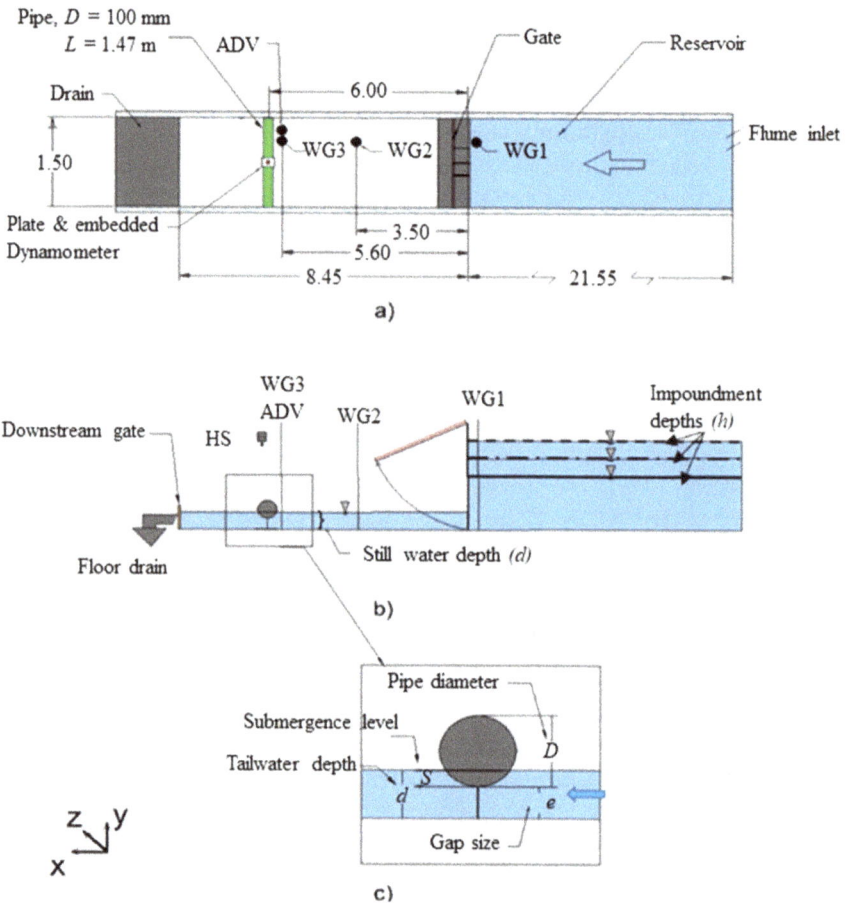

Figure 2. Experimental setting and instrument locations (unless otherwise specified, all dimensions are in m): (a) plan view, (b) side view, and (c) close view.

The second set of tests was conducted with the pipe installed inside the flume to study bore–structure interaction and to measure and analyze the forces exerted on the pipe due to the impacting bore. The pipe was filled with water in all the conducted tests. A detailed list of experimental parameters and results with respect to the hydrodynamics of this experimental program is presented in the companion paper [22].

3. Results and Discussion

3.1. Flow Hydrodynamics

The hydrodynamic properties of the generated dam-break waves were affected by the flume bed conditions, i.e., dry and wet bed conditions. The dry bed condition represents the first tsunami-like wave flowing on land, while the wet bed condition represents the following waves flowing on an existing layer of water remaining from the earlier wave attack. The flow hydrodynamics at the location of the pipe for dry and wet bed conditions are described below. Details of the flow hydrodynamics throughout the entire flume, as well as its change in the presence of the pipe, are discussed in more detail in the companion paper [22]. However, due to consistency and completeness of this work, a summary of the evolution of the surface elevation, velocity, and Froude number time-histories is presented below.

3.1.1. Dry Bed Condition

The flow characteristics at the pipe location were studied to analyze the forces exerted on the pipeline. Figure 3 illustrates the dry bed surge characteristics as the time-history of the water surface profile (wave gauge 3, WG3), the flow velocity, and the Froude number at the location of the pipe. The reference time for all experimental tests is the bore arrival time at WG3. There was a delay in velocity measurements using the ADV due to the air entrainment close to the wave arrival time, which corresponds to the zone with no data in the first few seconds in Figure 3b,c. Time-histories of the Froude numbers for the dry bed condition and different reservoir heights are shown in Figure 3. It should be noted that, for the impoundment depth of $h = 30$ cm, the water level increased and the flow velocity decreased more gradually compared to the cases with $h = 40$ cm and $h = 50$ cm. As shown in Figure 3c, the dry bed surge was supercritical, Fr > 1, throughout the studied time frame for all three impoundment depths. The Froude numbers remained almost constant in the case of $h = 30$ cm, whereas they gradually decreased for $h = 40$ cm and $h = 50$ cm.

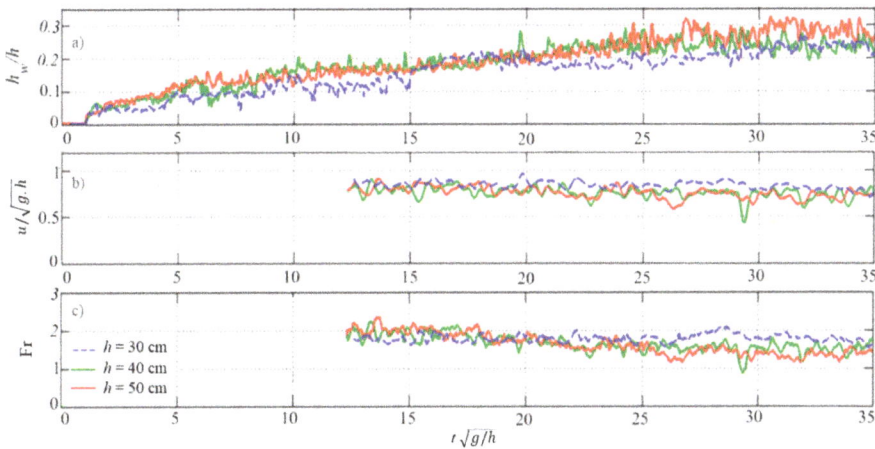

Figure 3. Dry bed surge characteristics at $x = 6.5$ for $h = 30$, 40, and 50 cm: (**a**) water surface profile (recorded by wave gauge 3, WG3), (**b**) flow velocity, and (**c**) Froude number.

3.1.2. Wet Bed Condition

Figure 4 shows the wet bed bore time-histories of the flow velocity and the derived Froude numbers for the cases where the wet bed condition was employed, for a constant impoundment depth, $h = 40$ cm, and for different d. The reference time for all the cases shown in Figure 4 is the wave arrival time to wave gauge WG3. The flow velocity data at the beginning of the bore propagation were considered invalid and were eliminated from Figure 4b–d. Results show a noticeable decrease in flow velocity (Figure 4b) and the estimated Froude number (Figure 4c), with an increase in the d/h ratio. This is because waves generated using a smaller pressure head (small difference between the impoundment depth and the downstream still water depth) result in slower flow velocities and smaller associated Froude numbers. Results from all the three tested impoundment depths, i.e., $h = 30$, 40, and 50 cm, show that, for $d/h > 0.3$, the flow was subcritical, while, for $d/h \leq 0.2$, the flow was supercritical. The d/h values around 0.2 resulted in a critical flow regime with the Froude number magnitude fluctuating around 1.0. A similar trend was observed for the case with $h = 30$ cm and $h = 50$ cm.

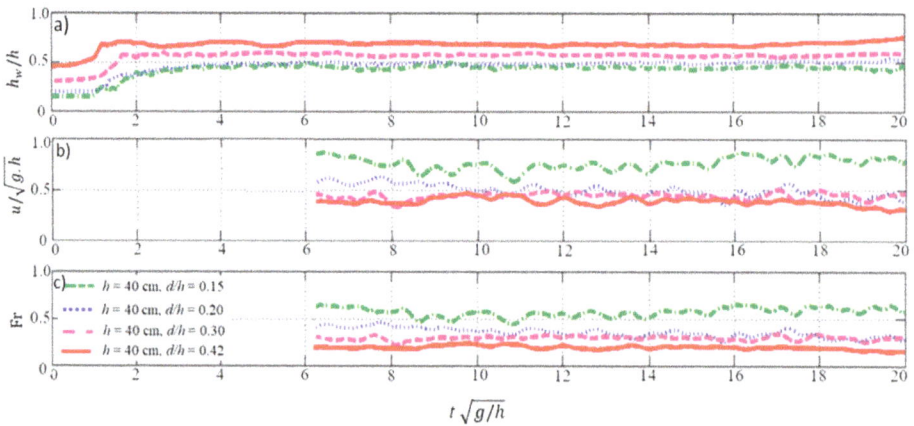

Figure 4. Wet bed bore characteristics at $x = 6.5$ for $h = 40$ cm, where different line types correspond to different d/h values: (**a**) water surface profile at WG3, (**b**) flow velocity, and (**c**) Froude number.

3.2. Drag and Lift Forces

Results from tests using different impoundment water depths and water depths downstream of the gate for different e/D values in wet and dry bed conditions are presented in this section. For clarity, all results are shown starting 1.0 s prior to the wave arrival time. The force magnitude in vertical (lift) and horizontal (drag) directions were set to zero shortly before the start of each test to eliminate the effect of the latent hydrostatic forces, as well as that of the pipe's own weight.

3.2.1. Force Time-History for the Dry Bed Condition

(a) Influence of pipe gap ratio (e/D)

Figure 5 shows the results from a test conducted with a wave generated by an impoundment depth $h = 40$ cm in the dry bed condition downstream of the gate. Figure 5a shows the water level time-history recorded by WG3 for the case with a gap ratio $e/D = 0.3$, while Figure 5b,c show the time-histories of the water level for $e/D = 0.6$ and 0.8, respectively. The schematic of the pipe with the corresponding distance from the bed is shown in Figure 5a–c for $e/D = 0.8$, 0.6, and 0.3, respectively, in order to visualize the level of the submergence and the full submergence time for each case. The water level for $e/D = 0.3$ rose rapidly immediately after the wave impact time, while it gradually increased for cases with gap ratios of $e/D = 0.6$ and $e/D = 0.8$. Figure 5d shows the measured drag force time-history for the three gap ratios. In Figure 5d, one can observe that, for $e/D = 0.3$, the drag force behavior was considerably different when compared to data obtained using other e/D ratios. For this particular value of $e/D = 0.3$, an impulse force at the wave arrival instant was recorded, whereas, for the other two gap ratios e/D, no such considerable force was recorded. In the cases with a large gap ratio ($e/D = 0.6$ and $e/D = 0.8$), water passed through the gap for the first few seconds of the wave propagation. However, in the case of a smaller gap size ($e/D = 0.3$), less water passed through the gap and, instead, a large portion of the incoming flow separated from the bed and surged on top of the pipe. This observation can also be explained by the sudden rise in the water level for $e/D = 0.3$ at $t = 1.5$ s (half a second after wave arrival).

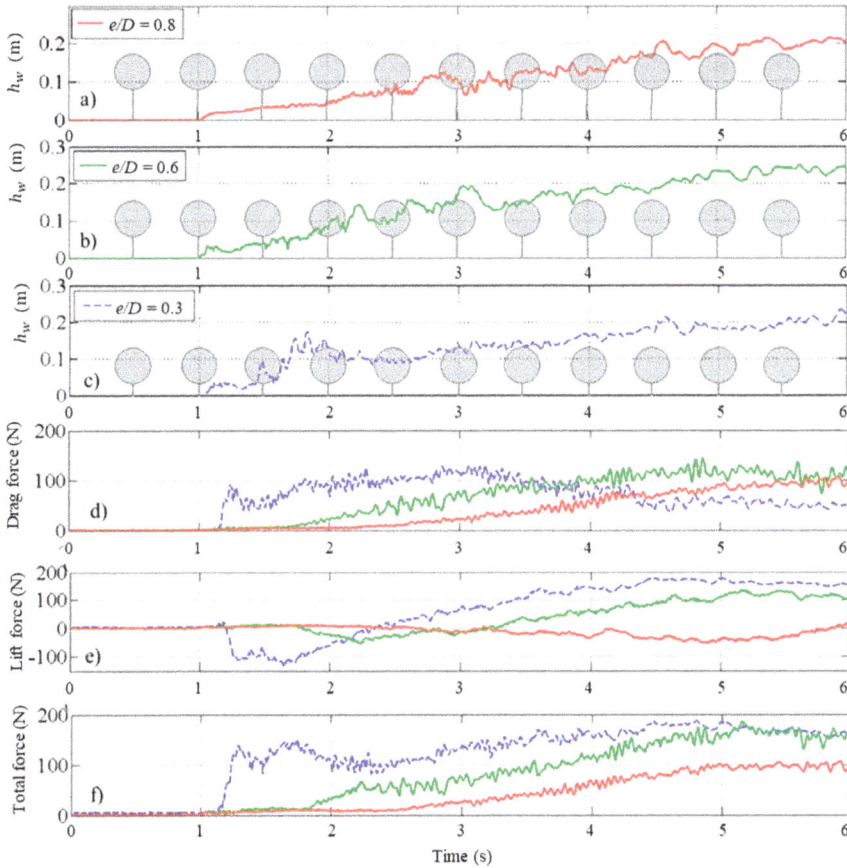

Figure 5. Time-history of the water level and the drag and lift force component measurements for the dry bed condition and an impoundment depth $h = 40$ cm: (**a**) water level with $e/D = 0.8$, (**b**) water level with $e/D = 0.6$, (**c**) water level with $e/D = 0.3$, (**d**) drag force time-history for $e/D = 0.3$, 0.6, and 0.8, (**e**) lift force time-history for $e/D = 0.3$, 0.6, and 0.8, and (**f**) total force for $e/D = 0.3$, 0.6, and 0.8.

Hence, the impulse force was considerably larger for the test with $e/D = 0.3$ while a longer rise time was observed for the force to reach its maximum value for the tests with $e/D = 0.6$ and $e/D = 0.8$. A smaller volume of the water passing through the gap and a shorter rise time for the tests with $e/D = 0.6$ when compared to that $e/D = 0.8$ resulted in higher water levels and larger drag forces at $e/D = 0.6$. For the test with $e/D = 0.3$, the water level reached the top edge of the pipe at $t = 3.16$ s. At this time, a surface roller formed right upstream of the pipe and started to propagate upstream, causing a considerable decrease in the drag force. The surface roller was also observed in the case of vertical structure obstructing a dam-break flow impacting a column by Reference [23]. For the other two gap ratios of $e/D = 0.6$ and 0.8, the observed surface roller was much smaller in size and the upstream propagation speed was considerably slower. Therefore, only in the test performed with $e/D = 0.3$ was a sudden decrease in the drag force observed. After $t = 4.4$ s, when the pipe was submerged in all the three gap sizes, the lowest drag force magnitude was associated with the gap ratio $e/D = 0.3$. This could be explained by the small distance of the pipe to the bottom, which caused the lower separation point to move downstream of the lower edge of the pipe while the upper separation point moved upstream of the upper edge. Such a separation point displacement by reduction of the gap ratio results in a larger wake size at the back of the pipe, which in turn results in a smaller drag force. This was observed and reported in previous studies [3,5,7,24]. Moreover, after full submergence, at smaller gap ratios (*e/D* < 0.3), vortex shedding suppression occurs due to the asymmetry of the vortices occurring at the top and bottom of the pipe. Therefore, von Kármán vortex shedding behind the cylinder suppresses and causes larger suction and smaller drag force exerted on the pipe.

The measured vertical (lift) force component time-history for the three gap ratios tested is shown in Figure 5e. This figure shows that the magnitude of the lift force increased considerably as the gap ratio decreased during the entire time-history. The variations observed in the time-history of the lift force occur mainly due to the displacement of the stagnation point. At larger gap sizes, the stagnation point moves away from the plane boundary which causes a considerable decrease in the lift force. A large gap ratio also diminishes the plane boundary effect, and the pressure distribution becomes symmetric. As a result, both the drag and lift forces decrease considerably. Moreover, at the beginning of the surge, the lift force at *e/D* = 0.8 was extremely small due to the longer duration it takes for the wave to reach the lower edge of the pipe. The downward-oriented lift force at the beginning of the bore surge at *e/D* = 0.3 can be explained with the large volume of water surging on top of the pipe immediately after the bore impact time, which pushes the pipe toward the flume bed. After the first impact, this downward vertical force was not observed in the tests when *e/D* = 0.6 and *e/D* = 0.8 as the flow passed through the gap. Figure 5f shows the total force which was calculated using the following equation:

$$F_T = \sqrt{F_y^2 + F_z^2}. \tag{4}$$

Results from Figure 5f indicate that the total force increased considerably as *e/D* decreased with the highest values recorded for the tests when *e/D* = 0.3. This is due to the larger drag and lift force observed at smaller gap ratios. The same trend in the force time-history was observed for the tests with the reservoir impoundment depths of *h* = 50 and 30 cm.

(b) Influence of the impoundment depth

Figure 6 shows the water level and the force components for different reservoir impoundment depths for the smallest *e/D* ratio used in the tests (*e/D* = 0.3). As expected, due to the larger initial head which translated into the dam-break wave with the highest flow depth, the reservoir depth *h* = 50 cm generated the largest force components magnitudes (Figure 6b). As discussed in the companion paper [22], higher impoundment depths lead to higher flow velocities and bore heights. This, in turn, increased the exerted drag force. For the impoundment depth *h* = 30 cm, the impulse force was considerably smaller compared to the two other larger impoundment depths. Waves generated using larger reservoir depths exhibited significantly steeper bore fronts, which impacted the pipe front face and caused a sudden water level rise and a larger impulse force at the time of bore impact. Figure 7 shows the more abrupt water level rise, which results in a larger impulse force in higher reservoir depths. For the cases with larger impoundment depth (*h* = 50 and 40 cm), large standing waves are generated in front of the pipe after the initial impact. After *t* = 4 s, the standing waves dissipated and started to move upstream toward the gate surface roller which resulted in a decrease in the drag force. In the case of the smallest impoundment depth *h* = 30 cm, standing waves and a returning surface roller were not observed. Therefore, the drag force increased gradually until reaching the maximum magnitude and remained then constant.

The lift force time-history in all three cases showed a downward peak which became larger with an increase in the reservoir impoundment depth from *h* = 30 to 50 cm. In Figure 6c, the negative peak of the vertical force (Fz) varied from −25 N to −50 N and to −150 N for water depths *h* = 30, 40, and 50 cm, respectively. This is due to the larger amount of water surging over the top of the pipe and pushing the pipe downward as the wave was larger. The lift force was generally larger for waves generated by the larger reservoir depths, mainly due to the higher flow velocity of the dam-break waves generated with higher impoundment depths. The total force magnitude, which is a function of the combination of the drag and lift forces, was maximum for waves generated using *h* = 50 cm (Figure 6d).

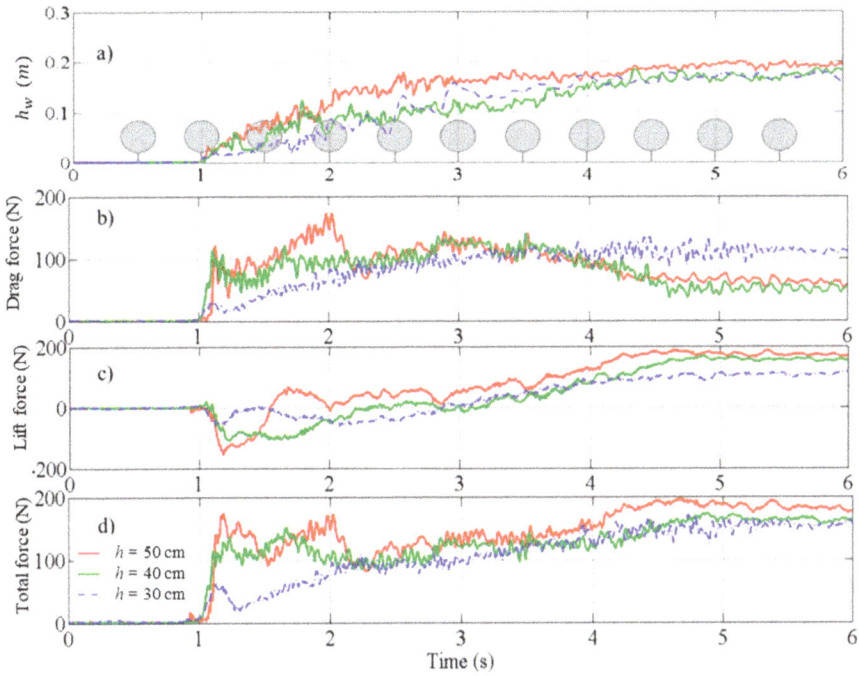

Figure 6. Time-history of the measured water level and force components for the dry bed condition for $e/D = 0.3$ and impoundment depths $h = 50$, 40, and 30 cm: (**a**) water level at WG3, (**b**) drag force time-history, (**c**) lift force time-history, and (**d**) total force time-history.

Figure 7. Bore impact on the pipe for three impoundment depths: (**a**) $h = 50$ cm, (**b**) $h = 40$ cm, and (**c**) $h = 30$ cm.

3.2.2. Force Components for the Wet Bed Condition

(a) Influence of pipe gap ratio (e/D)

Figure 8 shows hydrodynamic force components in the case of the wet bed condition for a wave generated by an impoundment depth $h = 40$ cm and a relative downstream water level $d/h = 0.2$. For the same initial forcing conditions, considerable changes in the hydrodynamic force time-history pattern can be observed at the time of bore impact when compared to the case with a dry bed condition; in the wet bed condition, the horizontal impulse force shown in Figure 8d for all three gap ratio values is significant compared to the case of a dry bed (Figure 5d). In the wet bed condition, the gap between the pipe and the boundary was filled with water. Therefore, unlike the dry bed case where the flow passed through the gap and gradually reached the pipe, the bore impacted the entire pipe cross-section at the bore arrival time, resulting in a large impulse force. Moreover, according to Yeh [25], the impulse force increases due to the impact of a steeper bore front onto the object in the path of the flow. Hence, the absence of a clear impulse force for the case of the dry bed condition can be attributed to the significantly milder slope of the bore as discussed in the companion paper [22]. From Figure 8d, it can be concluded that, for the wet bed condition, the drag force magnitude during the impact time does not change drastically by changing the gap ratio (e/D). Since, for the wet bed condition, the gap between the pipe and flume bed was partially or completely filled with water prior to the arrival of the bore, the effects of the gap ratio and boundary conditions, such as bed roughness, as well as separation and stagnation point, on the recorded force components are also decreased. Only the horizontally directed impulse force subsided in magnitude as the pipe was further submerged by the incoming bore and as the gap ratio decreased. This was due to an increase in the effective contact area of the pipe with the incoming surge by a lower initial level of submergence.

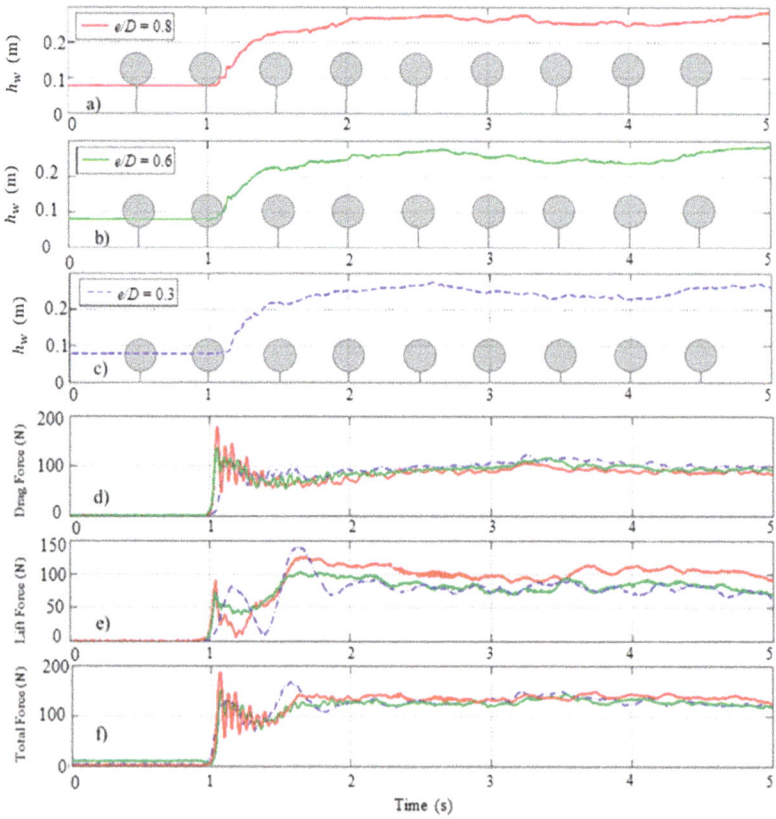

Figure 8. Time-history of the water level and force component measurements at WG1 for the wet bed condition, impoundment depth $h = 40$ cm, still water depth $d = 8$ cm, and $d/h = 0.2$: (**a**) water level for $e/D = 0.8$ and $S/D = 0$, (**b**) water level for $e/D = 0.6$ and $S/D = 0.2$, (**c**) water level for $e/D = 0.3$ and $S/D = 0.5$, (**d**) drag force time-history for $e/D = 0.3$, 0.6, and 0.8, (**e**) lift force time-history for $e/D = 0.3$, 0.6, and 0.8, and (**f**) total force for $e/D = 0.3$, 0.6, and 0.8.

Similar to the drag force time history for the wet bed condition, results of the lift force time-history in Figure 8e show a considerable initial impulse vertical force. A general trend for the lift force time-history $e/D = 0.6$ and 0.8 with a smaller initial submergence ratio ($S/D < 0.3$) was observed, where increasing e/D led to an increase in the impulse lift magnitude. However, for $e/D = 0.3$, with the initial level of submergence $S/D = 0.5$, the pipe was totally submerged right after the bore impacted the pipe, and the vertical force exhibited an oscillatory behavior with smoother changes in its time-history. The maximum peak of the lift force was observed for the smallest gap ratio ($e/D = 0.3$) due to the increased suction forces on the free-stream side of the pipe. No distinct difference between the results for different gap ratios was observed in the total force time-history (Figure 8f).

(b)   Influence of the wet bed still water depth d and submergence ratio S/D

Figure 9 shows the results for the water depth, the drag force, lift force, and the total force for different still water level depths downstream of the gate for the case of $e/D = 0.6$. The d/h ratio was adjusted by varying the still water depth d and using a constant impoundment depth h, which resulted in different levels of pipe submergence S/D as shown in Figure 9a.

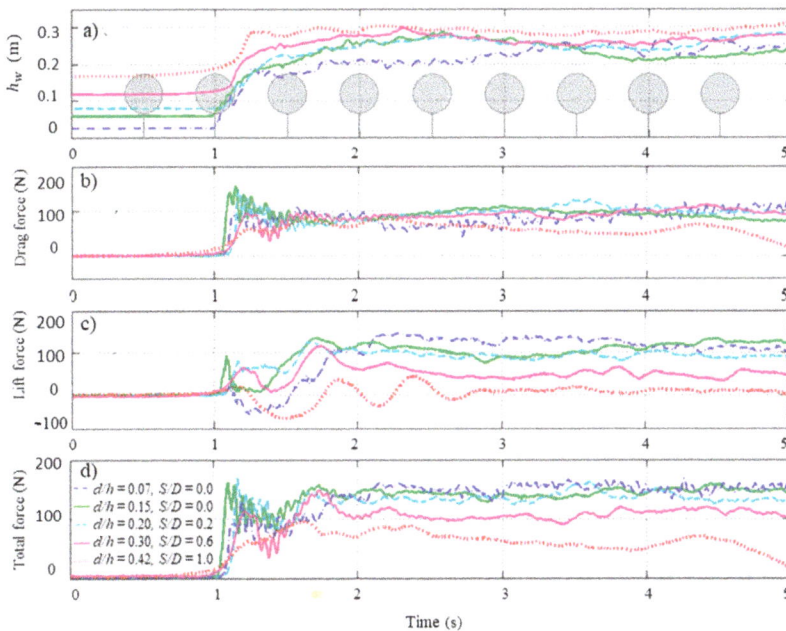

Figure 9. Time-history of the water level and force component measurements at WG3 for the wet bed condition, different d/h ratios ($d/h = 0.075, 0.15, 0.2, 0.3,$ and 0.42), and $e/D = 0.6$: (**a**) water level, (**b**) drag force time-history, (**c**) lift force time-history, and (**d**) total force.

The time-history of the drag force (Figure 9b) shows that. for $S/D = 0$ (the pipe not being submerged at the initial stage before opening the gate), the drag force initially exhibited an initial bore impact (impulse force) and further decreased to lower magnitudes afterward. When the initial still water level was below the bottom edge of the pipe ($d/h = 0.075$), the measured run-up force was larger compared to the impulse force just a few seconds after the bore impact. However, when the pipe was partially submerged with $S/D < 0.5$, the impulse force was the maximum one. For $S/D \geq 0.5$, a smaller impulse force was observed and the magnitude of the drag force remained constant as the pipe became fully submerged. In the case of a fully submerged pipe with $S/D = 1$, no considerable impulse force was recorded and a gradual drag force magnitude increase was observed, mainly due to an increase in the flow velocity. The run-up force was caused by the pipe obstructing the flow. Therefore, as the pipe was increasingly submerged, this force decreased. This occurred faster for the pipe with a larger initial level of submergence.

The time-history of the lift force shown in Figure 9c indicates a decrease in this force's magnitude with an increase in *d* and, as a result, an increase in the initial *S/D* ratio. Lowest values were observed for the case of the fully submerged pipe (*S/D* = 1). In this case, incoming flow only passed over the top of the pipe inducing small, mostly downward, vertical forces. The downward lift force at the time of the bore impact for the cases when *S/D* = 0 is due to the volume of water surging on top of the pipe at the time of bore impact. The rapidly surging water further pushed the pipe downward. Figure 9d shows the decrease in the total force time history with an increase in the *d/h* and *S/D* ratios. As discussed in the companion paper [22], a significant decrease in flow velocity was observed as the downstream still water level was increased. This explains the smaller force component magnitudes in larger *d/h* ratios. A similar trend in total force time-histories was observed for the cases when *e/D* = 0.3 and *e/D* = 0.8.

3.3. Force Coefficients

Force coefficients are used to determine the drag and lift forces exerted on a body placed within flow. Current design guidelines recommend force coefficient values for different bodies including horizontal cylinders exposed to steady flow conditions. For the case of unsteady flow conditions, force coefficients were suggested mostly for vertically oriented bluff bodies. For the first time, this research investigates the variation of force coefficients for horizontal cylinders placed in unsteady flow conditions (dam-break waves). Due to the unsteady nature of the dam-break flow, force coefficients are not constant but vary considerably over the duration of the flow–structure interaction [14]. In this study, the measured horizontal force incorporates both the hydrodynamic and hydrostatic components. Therefore, the term "resistance coefficient", rather than "drag coefficient", was used for the experimentally determined horizontal force coefficient, similar to studies by Gupta and Goyal [26] for steady flows around bridge piers, and Arnason [14] for tsunami impacts on vertical structures. The resistance coefficient for different pipe configurations, different impoundment depths, and different wet bed conditions downstream of the gate were calculated using

$$C_R = \frac{2F_H}{\rho L D u^2},$$ (5)

where F_H is the measured horizontal force during the experimental work, and *D* and *L* are the pipe diameter and pipe length, respectively. *u* is the depth average free-stream velocity as the average of ADV measurements at the highest water level, the location where the pipe center was later placed, and 0.03 m above the bed.

The lift coefficients were calculated using

$$C_L = \frac{2F_Z}{\rho L D u^2},$$ (6)

where F_z is the measured vertical force.

Figure 10 shows the calculated resistance coefficient values as a function of the Froude number for all experiments. The values of the maximum estimated resistance coefficients were used to define an upper envelope for the resistance coefficients, C_R. In the case of the dry bed condition, due to the higher flow velocities, the values of the resistance coefficients are found at the right side of the graph, corresponding to the larger Froude numbers which characterized this particular condition. Based on the results presented in Figure 10 the suggested C_R values vary between 1.0 and 3.5. Figure 10 shows that, for the case of supercritical flows (Fr > 1), C_R was almost constant, while it linearly increased with a decrease in Fr in the subcritical flow regime (Fr < 1). The black line in Figure 10 represents the upper envelope encompassing the entire range of experimental cases, from non-critical to supercritical flow conditions. As shown, it covers a variety of submergence conditions and gap widths, as well as both wet and dry bed conditions. The previous study by Li and Lin [26] on induced forces by waves and currents on horizontally submerged circular cylinders suggested drag coefficients in the range of $0.6 < C_D < 4$ for waves and $0.7 < C_D < 0.5$ for currents.

Similar to the resistance coefficient, the experimentally determined lift coefficients were plotted against the calculated Froude numbers (Figure 11). The maximum lift coefficients for all experimental cases were calculated by the authors to further propose an upper C_L envelope. This study suggests that C_L values for pipelines in unsteady flow conditions vary in the range of $0.5 \leq C_L < 3$. Similar to the resistance coefficient behavior, C_L remained almost constant in the supercritical flow region, while it increased as the value of Fr decreased in the subcritical flow region. Li and Lin [27] suggested lift coefficients in the range of $0.6 < C_L < 6$ for horizontal circular cylinders impacted by waves and $1.5 < C_L < 4$ for cylinders impacted by currents.

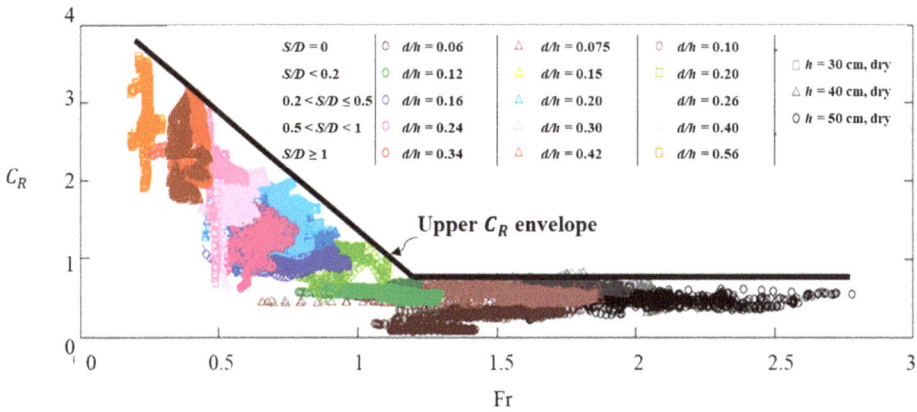

Figure 10. Calculated resistance coefficients versus Froude numbers for all the experimental cases tested in this study.

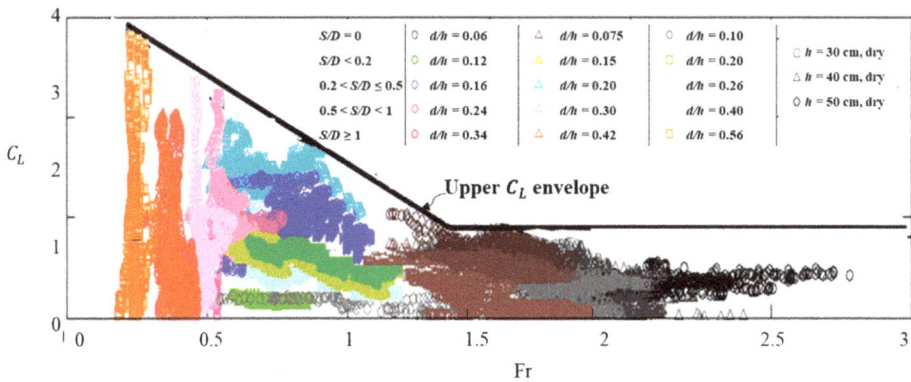

Figure 11. Maximum calculated lift coefficients versus Froude numbers for all the experiments conducted in this study.

4. Conclusions

A comprehensive experimental program was conducted to investigate the mechanisms of the extreme hydrodynamic loading exerted on a pipe due tsunami-like hydraulic bore flows. The time-histories of the hydrodynamic forces exerted on the pipe were measured and analyzed for different experimental conditions with respect to the pipe installation (distance/gap from the bed), as well as the flow characteristics (degree of submergence and hydraulic bores with different heights, as well as dry versus wet bed conditions over which the bore propagated).

Dam-break waves generated using higher reservoir impoundment depths propagating over the dry bed resulted in larger drag and lift forces due to considerably higher flow velocities compared to the case of the wet bed condition for the same relative impoundment depth.

Under the dry bed condition, the horizontal impulse force, observed at the initial impact of the bore, was considerably larger for smaller e/D, i.e., $e/D \leq 0.3$. The small gap space between the pipe and the bed led to a full impact of the steep bore front. However, after the complete submergence of the pipe, the magnitude of the horizontal force decreased due to vortex shedding suppression. The lift force time-history for the dry bed condition showed larger lift forces for smaller e/D ratios due to the asymmetry in the pressure distribution at the bottom and top of the pipe when this was placed close to the bed.

Changing the pipe distance to the bottom (gap ratio, e/D) for the wet bed condition did not significantly alter the force components. However, the impulse force recorded in the case of the wet bed condition was considerably larger for the same relative impoundment depth when compared to the same wave propagating under the dry bed condition. This is due to the impact of a steeper bore front on the entire pipe at the initial impact.

The effect of the downstream still water depth d was investigated, and it was concluded that, for d/h ratios corresponding to a smaller initial level of submergence S/D, the impulse force was considerably larger compared to the cases with larger submergence levels. For the case of the fully submerged pipe $S/D = 1$, no impulse force was observed due to the elimination of effective contact area of the pipe exposed to surge.

The pipe resistance coefficient exhibited lower values for the case of supercritical flow conditions (Froude number > 1.0), for both the dry and the wet bed conditions. This study suggested force coefficient values for various Froude numbers and several pipe configurations. The wide range of suggested force coefficients for various flow and pipe characteristics could be helpful for design purposes. The suggested resistance coefficients are in the range of $1 < C_R < 3.5$ and lift coefficient values in the range of $0.5 \leq C_L < 3$ for the experimental conditions investigated.

Author Contributions: B.G. developed the methodology and carried out the experiments. B.G. was also the main author responsible for the analysis of the data and writing the manuscript. J.S. assisted in conducting the experiments and contributed in reviewing and editing. I.N. and A.M. conceived the presented idea, supervised the work, and contributed in reviewing and editing. N.G. provided some of the instruments utilized in the study, assisted in the experimental set-up, and also contributed in reviewing and editing.

Funding: This research was funded by NSERC Discovery Grants held by Ioan Nistor, No. 210282 and Majid Mohammadian, No. 210717. Partial support for the study came through the Marie Curie International Outgoing Fellowship of Nils Goseberg within the 7th European Community Framework Program, No. 622214).

Acknowledgments: The authors are grateful to the University of Ottawa Hydraulic Laboratory Technician, Mark Lapointe, as well as to Adrian Simpalean and Derek Eden, graduate students at the University of Ottawa, for their assistance during the experimental work.

Conflicts of Interest: The authors declare no conflict of interest.

References

1. CEPA. *Pipeline Watercourse Management Program*; Canadian Energy Pipeline Association: Calgary, AB, Canada, 2013.
2. Sumer, B.M.; Fredsøe, J. *Hydrodynamics around Cylindrical Structures*; World Scientific Publishing Company: Singapore, 1997.
3. Lei, C.; Cheng, L.; Kavanagh, K. Re-examination of the effect of a plane boundary on force and vortex shedding of a circular cylinder. *J. Wind Eng. Ind. Aerod.* **1999**, *80*, 263–286. [CrossRef]
4. Aristodemo, F.; Tomasicchio, G.R.; Veltri, P. New model to determine forces at on-bottom slender pipelines. *Coast. Eng. J.* **2011**, *58*, 267–280. [CrossRef]

5.   Bearman, P.W.; Zdravkovich, M.M. Flow around a circular cylinder near a plane boundary. *J. Fluid Mech.* **1978**, *89*, 33–47. [CrossRef]

6.   Angrilli, F.; Bergamaschi, S.; Cossalter, V. Investigation of wall induced modifications to vortex shedding from a circular cylinder. *J. Fluids Eng.* **1982**, *104*, 518–522. [CrossRef]

7.   Zdravkovich, M.M. Forces on a circular cylinder near a plane wall. *Appl. Ocean Res.* **1985**, *7*, 197–201. [CrossRef]

8.   Buresti, G.; Lanciotti, A. Mean and fluctuating forces on a circular cylinder in cross-flow near a plane surface. *J. Wind Eng. Ind. Aerod.* **1992**, *41*, 639–650. [CrossRef]

9.   ASCE/SEI (ASCE/Structural Engineering Institute). *Minimum Design Loads and Associated Criteria for Buildings and Other Structures*; ASCE/SEI 7-16; ASCE/SEI: Reston, VA, USA, 2017; pp. 25–50.

10.   Li, C.W.; Lin, P.A. Numerical study of three-dimensional wave interaction with a square cylinder. *J. Ocean Eng.* **2001**, *28*, 1545–1555. [CrossRef]

11.   Sundar, V.; Vengatesan, V.; Anandkumar, G.; Schlenkhoff, A. Hydrodynamic coefficients for inclined cylinders. *Ocean Eng. J.* **1998**, *25*, 277–294. [CrossRef]

12.   Qi, Z.X.; Eames, I.; Johnson, E.A. Force acting on a square cylinder fixed in a free surface channel flow. *J. Fluid Mech.* **2014**, *756*, 716–727. [CrossRef]

13.   Nouri, Y.; Nistor, I.; Palermo, D. Experimental investigation of tsunami impact on free standing structures. *Coast. Eng. J.* **2010**, *52*, 43–70. [CrossRef]

14.   Arnason, H. Interactions between an Incident Bore and a Free-Standing Coastal Structure. Ph.D. Thesis, University of Washington, Seattle, WA, USA, 2005.

15.   Al-Faesly, T.; Palermo, D.; Nistor, I.; Cornett, A. Experimental modelling of extreme hydrodynamic forces on structural models. *Int. J. Prot. Struct.* **2012**, *3*, 477–506. [CrossRef]

16.   Bremm, G.C.; Goseberg, N.; Schlurmann, T.; Nistor, I. Long wave flow interaction with a single square structure on a sloping beach. *J. Mar. Sci. Eng.* **2015**, *3*, 821–844. [CrossRef]

17.   Foster, A.S.J.; Rossetto, T.; Allsop, W. An experimentally validated approach for evaluating tsunami inundation forces on rectangular buildings. *Coast. Eng. J.* **2017**, *128*, 44–57. [CrossRef]

18.   Aristodemo, F.; Tripepi, G.; Meringolo, D.D.; Veltri, P. Solitary wave-induced forces on horizontal circular cylinders: Laboratory experiments and SPH simulations. *Coast. Eng. J.* **2017**, *129*, 17–35. [CrossRef]

19.   FEMA P646. *Guidelines for the Design of Structures for Vertical Evacuation from Tsunamis*; Federal Emergency Management Agency: Washington, DC, USA, 2012.

20.   Ramsden, J.D. Tsunamis: Forces on a Vertical Wall Caused by Long Waves, Bores, and Surges on a Dry Bed. Ph.D. Dissertation, California Institute of Technology, Pasadena, XA, USA, 1993.

21.   Palermo, D.; Nistor, I.; Al-Faesly, T.; Cornett, A. Impact of tsunami forces on structures: The University of Ottawa experience. In Proceedings of the Fifth International Tsunami Symposium, Ispra, Italy, 3–5 September 2012.

22.   Ghodoosipour, B.; Stolle, J.; Nistor, I.; Mohammadian, A.; Goseberg, N. Experimental study on extreme hydrodynamic loading on pipelines. Part 1: Flow hydrodynamics. *J. Mar. Sci. Eng.* **2019**, *7*, 251. [CrossRef]

23.   St-Germain, P.; Nistor, I.; Townsend, R.; Shibayama, T. Smoothed-particle hydrodynamics numerical modelling of structures impacted by tsunami bores. *J. Waterw. Port Coast.* **2013**, *140*, 66–81. [CrossRef]

24.   Grass, A.J.; Raven, P.W.J.; Stuart, R.J.; Bray, J.A. The influence of boundary layer velocity gradients and bed proximity on vortex shedding from free spanning pipelines. *J. Energy Resour. ASME* **1984**, *106*, 70–78. [CrossRef]

25.   Yeh, H. Design tsunami forces for onshore structures. *J. Disaster Res.* **2007**, *2*, 531–536. [CrossRef]

26.   Gupta, V.P.; Goyal, S.C. Hydrodynamic forces on bridge piers. *J. Inst. Eng. Civ. Eng. Div.* **1975**, *56*, 12–16.

27.   Li, Y.; Lin, M. Hydrodynamic coefficients induced by waves and currents for submerged circular cylinder. *Procedia Eng.* **2010**, *4*, 253–261. [CrossRef]

© 2019 by the authors. Licensee MDPI, Basel, Switzerland. This article is an open access article distributed under the terms and conditions of the Creative Commons Attribution (CC BY) license (http://creativecommons.org/licenses/by/4.0/).

MDPI

St. Alban-Anlage 66

4052 Basel

Switzerland

Tel. +41 61 683 77 34

Fax +41 61 302 89 18

www.mdpi.com

Journal of Marine Science and Engineering Editorial Office

E-mail: jmse@mdpi.com

www.mdpi.com/journal/jmse

www.ingramcontent.com/pod-product-compliance
Lightning Source LLC
Chambersburg PA
CBHW051854210326
41597CB00033B/5889